CSR, Sustainability, Ethics & Governance

Series Editors

Samuel O. Idowu, London Metropolitan University, Calcutta House, London, United Kingdom

René Schmidpeter, Cologne Business School, Cologne, Germany

In recent years the discussion concerning the relation between business and society has made immense strides. This has in turn led to a broad academic and practical discussion on innovative management concepts, such as Corporate Social Responsibility, Corporate Governance and Sustainability Management. This series offers a comprehensive overview of the latest theoretical and empirical research and provides sound concepts for sustainable business strategies. In order to do so, it combines the insights of leading researchers and thinkers in the fields of management theory and the social sciences – and from all over the world, thus contributing to the interdisciplinary and intercultural discussion on the role of business in society. The underlying intention of this series is to help solve the world's most challenging problems by developing new management concepts that create value for business and society alike. In order to support those managers, researchers and students who are pursuing sustainable business approaches for our common future, the series offers them access to cutting-edge management approaches.

CSR, Sustainability, Ethics & Governance is accepted by the Norwegian Register for Scientific Journals, Series and Publishers, maintained and operated by the Norwegian Social Science Data Services (NSD)

Andrea Cardoni • Evgeniia Kiseleva

Sustainable Governance

Concept, Metrics and Contexts

Springer

Andrea Cardoni
Department of Economics
University of Perugia
Perugia, Italy

Evgeniia Kiseleva
Corporate Finance Center
HSE University
Moscow, Russia

ISSN 2196-7075 ISSN 2196-7083 (electronic)
CSR, Sustainability, Ethics & Governance
ISBN 978-3-031-37491-3 ISBN 978-3-031-37492-0 (eBook)
https://doi.org/10.1007/978-3-031-37492-0

This Springer imprint is published by the registered company Springer Nature Switzerland AG
The registered company address is: Gewerbestrasse 11, 6330 Cham, Switzerland

Preface

The role of Governance as a fundamental pillar of sustainability is now widely recognized and confirmed by its inclusion in the key ESG (Environmental, Social, and Governance) issues. Differently from the E and S, Governance has a more complex role, being a meta-construct over E, S, and ESG, and acting as a precondition and a result of the value creation in the long term.

In response to changing understanding of corporate governance in the context of ESG, research has developed the notion of sustainable corporate governance (Allais, Roucoules, & Reyes, 2017), that is, a system based on the integration of shareholders and all stakeholders' objectives, which safeguards the environment and collectivity at large.

Despite the uprising popularity of ESG agenda and high importance of "G" factor, the concept of sustainable corporate governance is in its early stages of development. The concept still lacks the profound theoretical definition, precise measurement system, and best practice examples of its implementation.

With an intent to bridge the abovementioned gap, the aim of the book *"Sustainable Governance: Concept, Metrics, and Contexts"* is to generalize the emerging concept of Sustainable Governance.

This work provides comprehensive theoretical foundations basing on stakeholder and institutional theories as well as context-specific theories for describing the concept in different settings and contexts, such as developed and developing economies, large listed companies and SMEs, family businesses.

The book systematizes components and operationalizes metrics of sustainable governance, being the first one in the field that builds a measurable sustainable governance concept. Additionally, the book makes it possible to reveal the value relevance of sustainable governance: the authors describe the dependence between sustainable corporate governance level and market valuations of the companies.

Apart from theoretical explanation of the concept, the authors use practical case studies to describe the effective integration of sustainable governance into corporate structures. The book contains an in-depth analysis of the best practice cases that illustrate not only the leadership practice in the development of some governance components, but also the financial value creation out of such leadership.

This publication provides students and researchers with a profound analysis of the emerging sustainable governance concept. Thorough literature review and wide usage of theories make this book insightful for scholars, while various case studies and analysis of financial market reactions build high relevance for companies and financial market participants as well as regulators and civil society.

The book is the result of a joint effort of the authors on the processes of conceptualization, planning, methodology, validation, data curation, writing, review and editing. The authors have read and agreed to the published version of the manuscript.

Perugia, Italy

Andrea Cardoni
Evgeniia Kiseleva

Reference

Allais, R., Roucoules, L., & Reyes, T. (2017). Governance maturity grid: A transition method for integrating sustainability into companies? *Journal of Cleaner Production, 140*, 213–226. https://doi.org/10.1016/j.jclepro.2016.02.069

Contents

About the Authors

Andrea Cardoni (PhD) is an Associate Professor of Business Economics at the Department of Economics at University of Perugia, Italy. He received his PhD in Management Control at the University of Florence and his current research interests include strategy and sustainability, management control, performance analysis, and business networking. Andrea has recently published his researches in journals like Business Strategy and the Environment, Business Process Management Journal, Measuring Business Excellence, Management Control and Sustainability.

He currently teaches Strategic Analysis and Business Valuation classes at post-graduate level in the Business Administration Course.

Andrea holds a former professional background as an auditor and consultant for one of the big four and he is motivated to reduce the theory–practice gap in strategy, governance, and management control for SMEs.

Evgeniia Kiseleva is a Doctor of Philosophy in the field of economics assigned by the University of Perugia, Italy, and HSE University employee (Russia). She specializes in research on sustainable value, sustainable governance, ESG ratings, and the connection between ESG scores and financial performance. Evgeniia has published research in such journals as Business Strategy and the Environment, Sustainability. Additionally, she taught classes on sustainability strategies and sustainable finance at the University of Perugia and HSE University.

Evgeniia has a background of work as a consultant on GRI and <IR> reports as well as corporate social responsibility strategies for large Russian companies. During the last years, Evgeniia initiated the launch of the local ESG rating and participated in the navigation of the project.

List of Figures

List of Tables

Part I

Chapter 1
Shifts in Corporate Governance Understanding

1.1 Agency Theory Interpretation of Corporate Governance: Why Did It Appear?

Corporate governance as a term appeared as early as the beginning of the twentieth century and is used to be interpreted from the point of view of agency problem that arose from the separation of the ownership and management in fast-growing corporations (Berle & Means, 1932).

Several factors served as a reason for corporate governance to appear. The size of the firms has been increasing, and the role of financial intermediaries and institutional investors was growing. Investors have met open markets with wider options for allocation of their funds, making investment decision more complex. Companies became more exposed to market forces and risks in the context of liberalization of financial and real markets and structural reforms, including price deregulation and increased competition (Claessens, 2006). To this end, the object under control of investor has become enormously large and not manageable by one person or even one corporate structure in case of institutional investors, what pushed investors to delegate part of their power. Since the separation between allocation and management of the capital occurred, holders of capital met such challenges as looseness of accounting standards in the context of changing corporate structure, the absence of a clear framework of how to control management, and competitive pressures both on companies and on auditors (*The Committee on the Financial Aspects of Corporate Governance and Gee and Co. Ltd.*, 1996).

The agency theory authors considered these factors and started to define the potential problems of the old business structures in the new contexts as well as generate theoretical concepts of how should business be organized under such new challenges. The agency problem is an essential element of the so-called contractual view of the firm, developed by Berle and Means (1932), Coase (1937), Jensen and Meckling (1976), Fama and Jensen (1983), and others as early as the 1930s. The essence of the agency problem is the separation of management and finance or—in

A. Cardoni, E. Kiseleva, *Sustainable Governance*, CSR, Sustainability, Ethics & Governance, https://doi.org/10.1007/978-3-031-37492-0_1

more standard terminology—of ownership and control (Shleifer & Vishny, 1997). According to basic agency theory (Hölmstrom, 1979; Jensen & Meckling, 1976; Ross, 1973), the relationship among shareholders and the board of directors is considered an agency relationship in which shareholders delegate the realization of corporate activities to the board of directors, whereas the actions or reactions of the boards of directors underline the information asymmetry between firm managers and financial markets (Eisenhardt, 1989). The agency theory of the firm suggests that when individuals engage in firm relationships, they are utility maximizers, self-seeking, and opportunistic, and, therefore, the governance system introduces mechanisms that align the interests of principals (owners) with those of their agents (the mangers) (Aguilera, 2005; Jensen & Meckling, 1976).

Why does the problem of agents cannot be controlled by the existing tools, such as contract between owner and management? Indeed, the financiers and the manager may sign a contract that specifies what the manager does with the funds and how the returns are divided between him and the holder of capital. Ideally, they would sign a full contract detailing what the manager does in each circumstance and how the profits are distributed. The problem is that most future contingencies are difficult to describe and predict, and as a result, complete contracts are technologically infeasible even if manager is fully motivated for raising funds and tries his best to develop a complete contract with multiple scenarios. Because of this problem of incomplete contracts, the manager and the shareholder have to deal with residual control rights, the rights to make decisions in circumstances not fully foreseen by the contract. To this end, the corporate governance scholars addressed exactly this problem by answering the question of how to allocate these residual control rights efficiently under the context of incomplete contracts (Shleifer & Vishny, 1997).

1.2 How Does the Agency Theory Governance Look Like?

According to agency theory authors, corporate governance concept served for dealing with agency problem with the aim to ensure profit maximization for shareholders in the context of incomplete contracts. In achieving the main aim, corporate governance was supposed to answer the following questions:

- How to motivate managers to return the incomes to the shareholders?
- How to make sure that managers will not seize the trusted capital or use it in the non-appropriate schemes?
- How do financial providers control the management of the company?

With the theoretical clearness on the issues, the great development of law regulation took place in the middle of the twentieth century, and boom of particular practical tools and detalization of the concept took place in the end of the twentieth century. According to Claessens (2006) and Shleifer and Vishny (1997), the main principles of the governance were formulated as follows:

- Protection of shareholders' rights (ownership of the asset, the right to receive profit, and the right to vote).
- Accountability (disclosure and transparency).
- Controllability (ability to control management activities).

The realization of these principles was supposed to be undertaken generally by two most common approaches to corporate governance, both of which rely on giving investors some power. The first is to empower investors by protecting them from expropriation by managers through laws. Examples of such mechanisms are the protection of minority rights and the legal prohibition of some managerial decision-making options. The second key approach is ownership by large investors (concentrated ownership): combining significant control rights with significant cash flow rights. Most corporate governance mechanisms used in the world—including large equity holdings, relationship banking, and even acquisitions—can be seen as examples of large investors exercising power (Shleifer & Vishny, 1997). Even better solution is to provide the manager ex ante highly conditional long-term incentives to align his interests with those of investors. Incentive contracts can take many forms, including stock ownership, stock options, or the threat of termination in low-income situations. The optimal incentive contract depends on the manager's risk aversion, the importance of the decision, and his ability to pay up front to have cash flow (Shleifer & Vishny, 1997).

In the end of twentieth century, practitioners joined the process of corporate governance development in companies: the first corporate governance codes appeared with explanation of the term as well as tools for its implementation.

The most influential practitioners' work in defining the field of corporate governance is the so-called Cadbury Report, entitled the Financial Aspects of Corporate Governance, which has been published by the Committee on the Financial Aspects of Corporate Governance chaired by Adrian Cadbury. The London Stock Exchange established the Cadbury Committee in 1991 to start work with mitigation of governance risks and errors, and report was published in draft form already in May 1992 with final version being published in December of the same year. The focus is on the code of best practice and the company's obligation to abide by it or explain to its shareholders why they have no obligation to do so. The recommendations and codes form the basis of the current corporate governance system in the UK and have proven to have an important impact on corporate governance developments around the world. A code of best practice was central to these, as was the demand that firms follow it or explain why they didn't. The proposals and the code laid the groundwork for the UK's current corporate governance structure and have had a significant impact on global corporate governance developments. The report's recommendations have been used, to varying degrees, to guide other regulations, such as those of the OECD, EU, US, World Bank, and others.

OECD Principles of Corporate Governance are the second most popular guidance that was issued in the last year of the century. The G20/OECD Principles of Corporate Governance were developed to assist policymakers in evaluating and improving the legal, regulatory, and institutional framework for corporate governance in order to enhance economic efficiency, long-term growth, and financial stability. The principles have established a global benchmark for policymakers, investors, corporations, and other stakeholders. They became as well one of the Financial Stability Board's Key Standards for Sound Financial Systems, and they served as the foundation for the World Bank's Reports on the Observance of Standards and Codes (ROSC) in the domain of corporate governance (OECD Publishing, 2015).

Similar to scientific interpretations, the codes depict corporate governance aim as to ensure that suppliers of capital get a return on their investment, where corporate governance itself is "the system by which companies are directed and controlled. Boards of directors are responsible for the governance of their companies. The shareholders' role in governance is to appoint the directors and the auditors and to satisfy themselves that an appropriate governance structure is in place. The responsibilities of the board include setting the company's strategic aims, providing the leadership to put them into effect, supervising the management of the business and reporting to shareholders on their stewardship. The board's actions are subject to laws, regulations and the shareholders in general meeting" (*The Committee on the Financial Aspects of Corporate Governance and Gee and Co. Ltd.*, 1996). In line with this document, the growth and evolution of the number of nonexecutive directors on boards, and the growing clarity on their roles and terms of "independence," was key for the development of corporate governance. This includes separating the role of CEO from that of nonexecutive chairman; this is to ensure that no individual has "unrestricted" decision-making power. The establishment of audit, compensation, and nominating committees composed of independent nonexecutives is also common and ideally ensures the correct use of incentives and a high degree of oversight of executive performance and decision-making. In addition to these internal controls, there are many external controls. First and foremost, here is the focus on improved "disclosure" and the "transparency" to achieve this, mostly related to financial performance, but more recently to social and environmental performance (Roberts, 2005).

The model described before was created for Anglo-American corporations and could be characterized by a combination of internal and external controls. Before 2001 and global scandals, there was growing confidence that this model was the best way to ensure effective governance and that corporate governance practices in other jurisdictions should and were beginning to converge on this model.

However, such a corporate governance understanding did not solve all of the problems of agency theory, as evidenced by the fact that during the wave of financial crises in 1998 in the Russian Federation, Asia, and Brazil, the corporate sector's behavior affected all economies, and corporate governance flaws jeopardized the global financial system's stability. Three years later, in 2001, corporate governance crises in the USA and Europe shattered investor trust, resulting in some of the greatest corporate failures in history (Claessens, 2006).

Corporate governance blind spots lead to shaky situations in the field with failures of major companies' governance structures. The Enron, WorldCom, and Tyco scandals emerged at this period, shaking trust in and complacency in the

Anglo-American paradigm and spurring more change. In the USA, the Sarbanes-Oxley Act can be interpreted as a near-perfect mirror of Enron's demise, implying a lack of trust in the self-regulatory capacity of both boards and markets by raising directors' criminal liability. The response in the UK has been not so sound, although the role for nonexecutive directors on boards increased as well as investor monitoring obligations in connection to voting, compensation, and activism (Roberts, 2005).

The Sarbanes-Oxley Act of 2002 is a federal law that creates broad audit practice and financial regulations for public companies. It was created to protect shareholders, employees, and the public from accounting errors and fraudulent financial practices. Auditors, bookkeepers, and company managers became accountable for the new set of rules. These rules amend and supplement several laws enforced by the Securities and Exchange Commission (SEC), including the Securities Exchange Act of 1934 and the Investment Advisers Act of 1940. The main areas of focus of the Sarbanes-Oxley Act are increase in criminal penalties, accounting regulations, new protection, and corporate responsibility—all of which are primarily designed to regulate financial reporting, internal auditing, and other business practices of public companies with certain regulations applicable to all businesses, including private businesses and nonprofit organizations.

Table 1.1 summarizes the agency theory interpretation of the corporate governance.

Table 1.1 Corporate governance understanding under the agency theory approach

Challenges	Business survival and development under the conditions of: • Fast-growing corporations with separation of ownership and control • Opportunism of management and incompleteness off contracts • Economic crises and corporate scandals (the crisis in Asia, Russia, and Brazil in 1998, the corporate scandals of the 2000s in the USA and Europe, financial crisis of 2008)
Questions to resolve	• How to motivate managers to return the incomes to the shareholders? • How to make sure that managers will not seize the trusted capital or use it in the non-appropriate schemes? • How do financial providers control the management of the company?
Aim	Help in gaining a high profit to shareholders by solving the agency problem
Definition	The system by which companies are directed and controlled
Principles	Protection of shareholders' rights (ownership of the asset, the right to receive profit, and the right to vote) Accountability (disclosure and transparency) Controllability (ability to control management activities)

Source: prepared by the author

1.3 Critics on Agency Theory Interpretation of Corporate Governance

Although scholars and practitioners were devoted to solving the agency problem, there was a particular critic of the idea in its core starting from the too narrow approach of agency theory, dynamism rejection and irrelevance for modern challenges in the world, and finishing with the neglect of the psychological aspects of human nature.

The first stream of criticism of agency theory presents it being too narrow and postulates that agency theory's corporate governance serves a shortsighted market model only, where society serves the interests of shareholders, leading to "competitive shortsightedness" and its consistent focus on near-term gains, market stress-induced earnings, stock prices, and other performance indicators. Agency theory limits stakeholders to one group—shareholders—making corporate governance unable to cope with external pressure from interest groups. Moreover, shareholders find it difficult and are reluctant to exercise all ownership-related responsibilities in public companies. At the same time, modern companies are influenced by a large number of stakeholders, including at least shareholders, lenders, customers, employees, suppliers, and management, who are often referred to as key stakeholders critical to the survival and success of the organization. In addition to these, companies have secondary but still important stakeholders such as local communities, media, courts, governments, interest groups, and the general public, i.e., society as a whole (Letza et al., 2008).

Narrowness of the agency theory concept relates not only to stakeholders but also to the results definition that takes only monetary forms in the concept, while there is no control over nonmonetary resources or outcomes. According to John Elkington, the "triple bottom line" will be the corporate necessity of the twenty-first century. Elkington contends in *Cannibals with Forks* that the three fundamental parts of the triple bottom line—social justice, economic prosperity, and environmental quality—will be the yardsticks against which corporate success will be assessed since corporate governance cannot be separated from social and other noneconomic circumstances and elements such as power, legislation, culture, social interactions, and institutional frameworks (Elkington, 1998; Letza et al., 2008).

Agency theory also limits the logic behind the personal and corporate choices made in economics. Corporate governance is based exclusively on economic logic, or rationality, and it assumes pure economic settings in which rational economic agents generate economic value separated from other social processes. Due to the reductionism, agency theory is comprehensible and easily operationalized but completely unable to explain and, even more obvious, predict complicated economic decisions with many-sided reasoning. Decision-making is a social process in which players (owners, managers, consumers, suppliers, and workers) negotiate based on their common and competing interests. Implications for agency theory, such as pure and free market in which resources are fairly divided and players have equal opportunities and abilities to use them, are also simplified and separated from real practice. In reality, resources and economic power are unevenly distributed and controlled by major decision-makers, decision-making is a multiple variable function, the number of choices and options are not particularly clear, and, finally, interactions between actors may play a great role in decision-making (Letza et al., 2008).

A second criticism has to do with the concept's lack of political dynamism. Political and economic forces are making visible influence on corporate governance and may drive companies to make governance more transparent with more emphasis on stakeholder inclusion. Powerful interest groups such as managers in the USA or banks in Germany may be interested in maintaining more power in hands of CEOs instead of shareholders or make preferences in rights for controlling shareholder even if it happens by the means of minority shareholders. Some authors have pointed out that corporate governance based on 1930s concepts is simply not ready for the new economic and political drivers. It does not take into account the dynamics inherent in design decisions: governance structures cannot be predesigned and generic but must emerge from a dynamic process in which agency decisions are not individually determined (Letza et al., 2008).

Elkington details which particular drivers companies were opt to meet in the late 1990s: author describes unprecedented challenges, or even revolution, that change the context of the business activities with seven drivers: markets, values, transparency, lifecycle technology, partnerships, time, and corporate governance. The first focuses on using market mechanisms rather than traditional command and control measures to improve performance in achieving sustainability goals. Next, sustainability challenge and technological innovation with improved eco-efficiency take place as a driver. Companies that seize the opportunity will be able to thrive on the coming wave of sustainability, while sustainability requirements may become entry barriers for lagging enterprises that did not react to those challenges. This revolution involves nonmonetary values becoming equal or even more significant than monetary values. Third, a transparency revolution makes a tremendous impact in highly communicative society, where company performance is increasingly judged and evaluated by outside observers. Lifecycle technology is the fourth revolution, involving the shift from perceiving company's product only at sale point to the whole supply chain focus, when child labor at contractor manufacture becomes unacceptable as child labor in the main office. Partnerships between business and activist groups are the fifth revolution. Campaigning organizations increasingly draw strength from society and define sustainability challenges as issues that need to be tackled together with business. The sixth revolution focuses on time: the time horizon of companies must shift from the short term to the long term. Last but not least, it is about power. The economy needs to strike a balance between shareholders and stakeholders, with a new definition of property rights to corporate assets. So the seventh revolution will revolve around corporate governance, which will develop the tools to act in the new environment in several revolutions in the field (Elkington, 1998).

Finally, the complex human nature is neglected in the corporate governance model developed under the roof of agency theory that presents an individual simply as the "fictitious atom" of an ideological representation—for example, within law or economics. The actual conduct of directors considered to be key factor in defining the effectiveness of the board. In contrast, Turley & Zaman (2007) found that the most significant impact of audit committees on corporate governance occurs outside of formal structures and processes, outside of official conduct. This is consistent with earlier observations that corporate governance is a social process and should therefore also be reviewed from a social perspective. From an institutional point of view, corporate governance is a secondary institution that is influenced by larger institutions and social forces as well. Understanding these forces will help to understand that corporate governance is not a mechanism for dealing with agency issues, but rather a political and social tool (Roberts, 2005; Yusof, 2016).

Considering the deep nature of the problems associated with the agency theory, some scholars state that the problem lies in the core of the concept itself. The tools of corporate governance—a mixture of increased independent monitoring, sharper sanctions, and more appropriately targeted incentives that avoid "reward for failure"—may be seen not as the solution but rather as the source of the governance problem (Roberts, 2005).

1.4 The Answer to the Agency Theory Problems: Stakeholder Theory?

Then what is the solution for effective corporate governance? And what factors should corporate governance include into its responsibility in order to be ready for new world challenges, oriented toward wider list of stakeholders and enlarged performance definition with triple-bottom line factors in mind and, finally, applicable to real person with its complex behavior patterns in economy and everyday life?

Overview 1.1 Knowledge-based enterprise

In attempts to answer the questions about effective corporate governance, alternative to agency theory approach, Zingales et al. (Rajan & Zingales, 2000) discuss the issues of corporate governance from the point of view of knowledge-based perspective in the research. It is the wave of corporate governance research that perceives the firm as a bundle of knowledge and puts an emphasis on human capital.

Authors start from the fact that the world is changing: the boarders are opening for the international trade and international move of human resources, innovations, and intangible assets that start to play more important role. These factors stimulate the change in the organizations and their boundaries. Zingales et al. illustrate the changes by the examples of traditional enterprises and the new enterprises.

(continued)

Overview 1.1 (continued)

Traditional firm is represented as a strict hierarchical structure, vertically integrated giant, oligopolist, or monopolist in the local market. The boundaries of the firm are well determined by legal criteria: the ownership determines the boundaries that cover upstream and downstream value chain units. The inanimate capital resources were of paramount importance; thus, the power was in the hands of owners, and the surplus gained from the resources was accumulated at the top. The reallocation of the resources and surplus between organization's units was undertaken with the help of fiat, not with the help of market forces.

However, the openness of the world and the increased role of intangible assets lead to the dramatic changes in the governance structure of the firm: vertically integrated giants started to break down while the "new enterprise" appeared. The "new enterprise" is more specialized in the value chain—it is no more vertically integrated. The outsource is becoming more effective than keeping not productive units inside the organization. What is more important, the role of capital assets decreases with the development of new technologies. The role of owners of these assets is decreasing, and the economic boundaries of the firm spread far beyond the legal definition. As a result, the shareholders' power covers not all the firm, and new powerful stakeholders appear.

With the salient changes in the organization, the main question of the corporate governance is also changed.

The traditional enterprise structure was well defined, and legal boundaries were the same as the economic boundaries. Thus, the main question was how to efficiently allocate vital resources. Moreover, the large size of the organization led to the fact that operational control was delegated to outside managers. Such separation between ownership and control led, in turn, to the agency problem between shareholders and top managers. The shareholders' value maximization and incentives for manager to increase this value were the main issues of concern.

The corporate governance of the new enterprise has completely different tasks. As far as economic boundaries of the firms are wider than legal ones, the question of keeping the power arises. New powerful stakeholders appear and it should be taken into consideration.

Zingales et al. claim that the main new stakeholder is an employee. Authors explain that the employee is the owner of the knowledge, innovations, and intangible assets that are of paramount importance in the new enterprise. Thus, another one question appears how to retain employees and attract them in common value generation inside the enterprise. Zingales et al. state that there are two ways to ensure this. Firstly, the company and the employee should be complementary to each other so that cocreated value should be greater than the value created by the employee or by the company alone. Additionally, the

(continued)

Overview 1.1 (continued)
company may provide an access to some valuable resources to the employee what will be a motivator for him to stay inside the organization.

The work by Zingales et al. also describes the changes in value creation process: since main holders of the power are the employees, the value should be cocreated by company and its employees.

Other authors postulate that not only employees but also government, citizens, suppliers, environmental organizations, and other stakeholders start to influence value creation processes (Freeman, 1984). Thus, one may suppose that in order to ensure stable value creation, all the stakeholders should be attracted to the value creation and value distribution processes in view of the fact that cooperation enables greater generation of value. Thus, the question of the corporate governance may become not only how to attract employees to value cocreation but also how to attract all the essential stakeholders. This idea corresponds to the Hart and Milstein concept of sustainable value (Hart & Milstein, 2003) and Porter's idea of shared value (Porter & Kramer, 2011) where value concept is much wider than just shareholder value. Such wider understanding of value creation may lead to better understanding of the role of the corporate governance.

Legitimacy theory is often stated as a next step in understanding corporate governance, where different stakeholders are considered instead of solely shareholders and investors. Legitimacy theory is based on the common sense that the company operates inside the value system of the society (Singh et al., 1986). Central to legitimacy theory is the concept of a social contract between organizations and members of a society that allows companies to operate as long as they fit the requirements imposed by society to the social and environmental impact of the business. If the society feels that the company does not fit its norms, it can imply sanctions in the form of restrictions on the firm's operations, resources, and demand for its product (Lindblom, 1994). Thus, the survival of an organization depends on whether it behaves in socially acceptable ways. The wish of the company to legitimize its activities inside the system it operates explains the motives of corporate managers to disclose the company's corporate governance and ESG (environmental, social, and governance issues) and implement social and environmental projects, stakeholder communication tools, diverse governance structure, compliance system, etc. (Deegan, 2002). In this case, the sustainable governance becomes a tool of legitimization of company's activities. Governance of ESG issues becomes an entrance barrier to the market for the companies where the legitimization effect is very high.

Legitimacy theory introduces the notion of some other risks existing apart from the financial risks and importance of ESG risk management and building a trustful environment around the company and loyal partners. However, stakeholder theory brought wider perception of interested parties, where they may bring not only risks but opportunities, making sustainable governance not only a cost-saving mechanism but also a competitive advantage.

Stakeholder theory interpretation of corporate governance is gaining momentum as an alternative point of view on corporates that has a potential to overcome the abovestated drawbacks of previously commonly accepted agency theory and largely enrich legitimacy theory that makes emphasis only on normative pressures and risks as a motivation for development.

The discussion over main stakeholder(s) of the company started as early as 1932 with Berle article "For Whom Corporate Managers Are Trustees: A Note." Being a pioneer in the field, Berle doubted the correctness of a historically formulated rule that corporate governance mission is to run an enterprise for the benefit of its security holders. He explained that other groups, especially workers, make their claims from time to time, and these claims are becoming stronger and often associated as a cost. If these fees are neglected, shareholders earn illusory additional profits, but company is subject to high social risks of losing reputation or employees loyalty (Berle, 1932).

Stakeholder theory has grown in popularity in recent years, to the point that it can now be considered a subject of academia, although one with a wide range of topics. Its popularity is likely due to a number of factors, including an increasingly complex and interconnected external environment that stakeholder theory is well suited to address, recognition among business scholars and managers that placing too much emphasis on short-term financial returns has resulted in unfavorable outcomes for businesses and society, numerous high-profile business scandals that have raised public awareness of ethical issues, and a global sustainability movement (Phillips et al., 2019).

Stakeholder theory is usually associated with R. E. Freeman, an American philosopher and professor of business administration at the Darden School of the University of Virginia, and his popular work *Strategic Management: A Stakeholder Approach* of 1984. According to the author, managerial capitalism and the assumption that managers have a duty to investors are replaced with the idea that managers have a fiduciary relationship with stakeholders, where stakeholders are persons or organizations that may be interested in company's activities since they are impacted by the enterprise operations. Suppliers, customers, workers, investors, local community, as well as management in its function may be considered as key stakeholders that want to and have a social right to the right not to be viewed not as a means to a goal but as participants in choosing the firm's future path in which they have an interest (Freeman, 2016).

The core argument of stakeholder theory is that a broader purpose of the corporation is fairer and socially efficient than one focused just on shareholder profit. Other organizations and individuals with a long-term relationship with the business and hence a "stake" in its long-term performance, such as workers, suppliers, customers, and managers, should be recognized. Instead of maximizing shareholder returns, corporate governance should be aimed at maximizing the corporation's overall wealth generated or value. Such value generated is usually called sustainable value, cocreated, shared, or simply stakeholder value, and means monetary and nonmonetary impacts generated by companies together with its stakeholders and for these stakeholders. That is, an organization creates value by providing utility to a wide range of stakeholders. For example, customers and clients receive utility as they make use of the products and services of the firm; employees receive stable salary with social security programs, education, and growth opportunities; communities may receive positive impacts by the projects that are undertaken by company together with local charitable organizations; and so forth. The vital characteristic of such behavior of the firm under stakeholder theory assumptions is that it is not an imposed obligatory scheme from above to comply but a voluntary behavior that is integrated into corporate structures on a cultural level throughout all organization (Letza et al., 2008).

References

Aguilera, R. V. (2005). Corporate governance and director accountability: An institutional comparative perspective. *British Journal of Management, 16*(s1), 39–53. https://doi.org/10.1111/j.1467-8551.2005.00446.x

Berle, A. A. (1932). For whom corporate managers are trustees: A note. *Harvard Law Review, 45*(8), 1365–1372. https://doi.org/10.2307/1331920

Berle, A. A., & Means, G. (1932). *The modern corporation and private property.* Commerce Clearing House. https://edisciplinas.usp.br/pluginfile.php/106085/mod_resource/content/1/DCO0318_Aula_0_-_Berle__Means.pdf

Coase, R. H. (1937). The nature of the firm. *Economica, 4*(16), 386–405. https://doi.org/10.1111/j.1468-0335.1937.tb00002.x

Claessens, S. (2006). Corporate governance and development. *The World Bank Research Observer, 21*(1), 91–122. https://doi.org/10.1093/wbro/lkj004

Deegan, C. (2002). Introduction: The legitimising effect of social and environmental disclosures – A theoretical foundation. *Accounting, Auditing & Accountability Journal, 15*(3), 282–311. https://doi.org/10.1108/09513570210435852

Eisenhardt, K. M. (1989). Agency theory: An assessment and review. *Academy of Management Review, 14*(1), 57–74. https://doi.org/10.5465/amr.1989.4279003

Elkington, J. (1998). *Cannibals with forks: The triple bottom line of 21st century business.* New Society Publishers.

Fama, E. F., & Jensen, M. C. (1983). Separation of ownership and control. *The Journal of Law & Economics, 26*(2), 301–325. https://www.jstor.org/stable/725104

Freeman, R. (1984). *Strategic management: A stakeholder approach.* Cambridge University Press.

Freeman, R. E. (2016). A stakeholder theory of the modern corporation. In *A stakeholder theory of the modern corporation* (pp. 125–138). University of Toronto Press. https://doi.org/10.3138/9781442673496-009.

Hart, S. L., & Milstein, M. B. (2003). Creating sustainable value. *Academy of Management Perspectives, 17*(2), 56–67. https://doi.org/10.5465/ame.2003.10025194

Hölmstrom, B. (1979). Moral hazard and observability. *The Bell Journal of Economics, 10*(1), 74–91. JSTOR. https://doi.org/10.2307/3003320

Jensen, M. C., & Meckling, W. H. (1976). Theory of the firm: Managerial behavior, agency costs and ownership structure. *Journal of Financial Economics, 3*(4), 305–360. https://doi.org/10.1016/0304-405X(76)90026-X

Letza, S., Kirkbride, J., Sun, X., & Smallman, C. (2008). Corporate governance theorising: Limits, critics and alternatives. *International Journal of Law and Management, 50*(1), 17–32. https://doi.org/10.1108/03090550810852086

Lindblom, C. (1994). The implications of organizational legitimacy for corporate social performance and disclosure. *Critical Perspectives on Accounting Conference*, New York, 1994. https://cir.nii.ac.jp/crid/1571135650849054080

OECD Publishing. (2015). *G20/OECD Principles of Corporate Governance.* Retrieved August 7, 2022, from https://doi.org/10.1787/9789264236882-en

Phillips, R., Barney, J., Freeman, R., & Harrison, J. (2019). Stakeholder theory. *The Cambridge Handbook of Stakeholder Theory*, 1–16. https://scholarship.richmond.edu/management-faculty-publications/107

Porter, M. E., & Kramer, M. R. (2011). The big idea: Creating shared value. How to reinvent capitalism. *Harvard Business Review, 89*(1/2), 62–77.

Rajan, R. G., & Zingales, L. (2000). *The governance of the new enterprise.* https://papers.ssrn.com/sol3/papers.cfm?abstract_id=245587

Roberts, J. (2005). Agency theory, ethics and corporate governance. In C. R. Lehman, T. Tinker, B. Merino, & M. Neimark (Eds.), *Corporate governance: Does any size fit?* (Vol. 11, pp. 249–269). Emerald Group Publishing Limited. https://doi.org/10.1016/S1041-7060(05)11011-6

Ross, S. A. (1973). The economic theory of agency: The principal's problem. *The American Economic Review, 63*(2), 134–139. JSTOR.

Singh, J. V., Tucker, D. J., & House, R. J. (1986). Organizational legitimacy and the liability of newness. *Administrative Science Quarterly, 31*(2), 171–193. https://doi.org/10.2307/2392787

Shleifer, A., & Vishny, R. W. (1997). A survey of corporate governance. *The Journal of Finance, 52*(2), 737–783. https://doi.org/10.1111/j.1540-6261.1997.tb04820.x

The Committee on the Financial Aspects of Corporate Governance and Gee and Co. Ltd. (1996). *Report of the Committee on the Financial Aspects of Corporate Governance* (Reprinted). Retrieved August 7, 2022, from https://www.icaew.com/-/media/corporate/files/library/subjects/corporate-governance/financial-aspects-of-corporate-governance.ashx?la=en

Turley, S., & Zaman, M. (2007). Audit committee effectiveness: Informal processes and behavioural effects. *Accounting, Auditing & Accountability Journal, 20*(5), 765–788. https://doi.org/10.1108/09513570710779036

Yusof, N. Z. M. (2016). Context matters: A critique of agency theory in corporate governance research in emerging countries. *International Journal of Economics and Financial Issues, 6*(7S), 154–158. https://www.econjournals.com/index.php/ijefi/article/view/3599

Chapter 2
ESG or Sustainability Agenda and Its Influence on the Role of Corporate Governance

2.1 Sustainability Agenda

Climate change as a result of rising temperature on the planet, problems of income inequality, poverty and lack of resources, as well as financial crises have attracted the attention of the corporate world and the scientific community to the sustainability challenges:

The modern sustainability challenges that companies need to meet are wide (United Nations Development Programme, 2022):

- The expected change in the average temperature of the Earth's surface in 2081–2100 compared to 1986–2005 is 2–4–6 °C. To ensure carbon neutrality in 2050, it will take from $3 to $5.7 trillion per year.
- Disasters and the effects of climate change have displaced more people than ever before—on average 14 million people annually.
- Approximately 700 million people still live on less than US$1.90 per day, a total of 1.3 billion people are multidimensionally poor including a disproportionate number of women and people with disabilities, and 80% of humanity lives on less than US$10 per day.
- Conflicts, sectarian strife, and political instability are on the rise, and more than 1.6 billion people live in fragile or conflict-affected settings.

In order to overcome the difficulties, the United Nations developed the well-known Sustainable Development Goals, a set of 17 interrelated goals developed in 2015 by the UN General Assembly as a plan to achieve a better and more sustainable future for all (United Nations, 2015).

The measurable part of sustainability agenda that is applicable to the corporate world is ESG—the agenda in the field of environmental and social spheres and sustainable corporate governance.

Investors, banks, rating agencies, regulators, the scientific community, and other stakeholders have begun to revise the criteria for business performance, including

A. Cardoni, E. Kiseleva, *Sustainable Governance*, CSR, Sustainability, Ethics & Governance, https://doi.org/10.1007/978-3-031-37492-0_2

not only financial success but also the ability to solve more complex issues—to respond to challenges in the field of ESG.

Traditionally the social responsibility focus has been associated with the public pension funds and union pension funds (Erhemjamts & Huang, 2019). Later, larger scale of investors began to take into account environmental and social risks, integrate ESG aspects into their activities, and actively discuss topics related to ESG aspects. Under the influence of the COVID-19 pandemic and the subsequent social and economic crisis, attention to the topic of sustainable development has intensified and led to an abundance of ESG initiatives from market participants.

Investors

More and more investors are including ESG criteria in investment valuation models:

- ≈$2.4 trillion—assets under management in funds that self-identify as having ESG mandates and they
  - Tripled in 2010–2020.
  - Increased by 50% in 2019.
  - Doubled in 2020 (Scatigna et al., 2021).
- ≈36%—ESG assets (including ESG funds) share out of total professionally managed assets (Aramonte & Zabai, 2021).
- ESG investments make up almost 18% of foreign financing for emerging markets excluding China (Gautam et al., 2022).

Companies

Companies publish ESG data and reconsider the aims of the business.

Large European companies have been publishing ESG data according to the Non-Financial Reporting Directive (NFRD) since its publication in 2014 and according to more comprehensive Corporate Sustainability Reporting Directive (CSRD) that is issued by the EU Commission in 2023. Small and medium enterprises (SMEs) in Europe with listing at stock exchanges are also exposed to the new regulation, while the EU Commission discusses a separate nonfinancial reporting standard for all SMEs.

Companies are using variety of standards for nonfinancial reporting. Global Reporting Initiative (GRI) standards are the most cited internationally in sustainability reporting (KPMG, 2020). Integrated Reporting Framework (<IR>) as well as Sustainability Accounting Standards Board (SASB) are widely used and expected to be included into IFRS Sustainability Reporting Standards (SASB, 2021).

Apart from becoming transparent, business is reconsidering the aims of the operation. Business Roundtable, the organization, which includes the heads of the largest US companies and which once defined the company's goal as protecting the interests of shareholders, has recently changed this goal to include obligations to all interested parties (Business Roundtable, 2022).

Regulators and Standard Setters

Government agencies and international nongovernmental organizations issue documents regulating the behavior of market participants in the context of ESG. Most of the documents are voluntary, but with the potential to be translated into mandatory rules:

- In 2019, the European Parliament issued Sustainable Finance Disclosure Regulation (SFDR) to increase the transparency of financial institutions in the ESG areas (The European Parliament And The Council Of The European Union, 2019).
- In 2021, the International Organization of Securities Commissions (IOSCO) issued ESG Ratings and Data Products Providers Consultation Report (IOSCO, 2021).
- Meanwhile, EU launched the Sustainable Corporate Governance Initiative and articulated plans to issue a report on the expected effect of the initiative for the development of sustainable corporate governance. France, the Netherlands, and Italy already have legislative regulations governing the development of elements of ethical corporate governance (European Commission, 2019).
- In November 2021, the IFRS published a revised Constitution and a Feedback Statement that responds to the feedback from Exposure Draft Proposed Targeted Amendments to the IFRS Foundation Constitution to Accommodate an International Sustainability Standards Board to Set IFRS Sustainability Standards (SASB, 2021).
- In 2023, the final Corporate Sustainability Reporting Directive (CSRD) by the European Parliament came into force, which amends the current Non-Financial Reporting Directive (NFRD). It requires almost 50,000 companies across EU to disclose information on business model and strategy, policies, risks, targets, and due diligence in relation to ESG issues: employee aspects, respect for human rights, anticorruption and bribery, gender and age diversity, educational and professional backgrounds of the board, etc. (The European Parliament and the Council of the European Union, 2022).

Scientific Community

Universities are paying more and more attention to the analysis of topics related to the ESG agenda with boost of interest to the topic in 2020 and 2021 (Fig. 2.1).

2.2 New Role of Business under ESG Challenges

Considering the uprising interest in ESG integration, the notion of business is changing as well as the value perception (Freeman et al., 2018). Before, an organization was considered as a black box that uses resources and generates economic profits for shareholders. Today, attention has been turned toward sustainable value

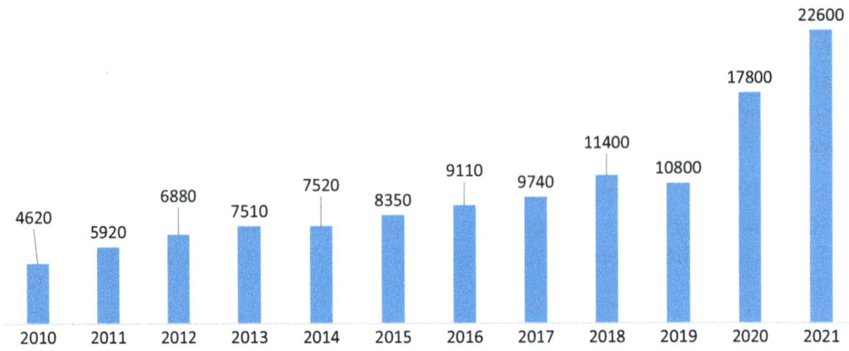

Fig. 2.1 Number of articles using the abbreviation ESG, Google Scholar. Source: constructed by the author

creation or cocreation with stakeholders over a longer period of time (Chandler, 2016).

Speaking about classical capitalistic approach, the delivery of economic value to shareholders is a dominant business concern. However, pure economic value creation may often destroy the value for legitimate stakeholders (Brennan & Tennant, 2018).

Maximization of profit as the only one aim of business is becoming an obsolete idea. Instead, companies tend to optimize value over the long term with reorientation of efforts toward the creation of sustainable value (Chandler, 2016). Accepting sustainable value creation as the pivotal element of business will allow concepts of CSR, sustainability, and the stakeholder approach to find their places, at strategic and managerial levels at the same time (Wheeler et al., 2003).

The increased importance of stakeholders in business processes changes the perception of the aims of the business and of the beneficiaries of value created. The attention is attracted to the cocreation of value with stakeholders, in collaboration and win–win partnerships with further distribution of the value to the key stakeholders, including but not limited to shareholders (Tantalo & Priem, 2016).

From the theoretical point of view, the understanding of business moves from the resource-based perception and agency theory interpretation of corporate governance to the stakeholder theory. The first examined a firm's competitive advantage emerging from unique resources under condition of the limited amount of strategic resources. Agency theory put an emphasis on solving the problem of the conflict between managers and shareholders for the purposes of profit maximization. On the other hand, stakeholder theory's distinctive attribute is based on the strategic stakeholder communication and integration of stakeholders in the value creation processes, where value is created with stakeholders and for stakeholders (Freeman et al., 2018, 2021).

Definition 2.1 Resource-based theory—theory of the nature, behavior, and performance of firms, in which the unit of analysis is a resource or capability that only this firm owns, controls, or has an access to.

Definition 2.2 Agency theory—a theory that explains the important relationships between principals and their relative agent. In case of corporate governance, between shareholder and manager of the company.

Definition 2.3 Stakeholder theory—theory that refers to the ethical concept and considers main outcome of business decisions to be generated together with interested parties and shared with them. Such interested parties are usually called stakeholders and include shareholders, employees, financers, government, customers, suppliers, society, and even environment as a nonhuman stakeholder.

In relation to the resource-based view, stakeholder theory adds valuable opportunities that can be used by the companies in meeting sustainability challenges of twenty-first century. Out of such possibilities, stakeholder theory adds normativity context when norms, values, and ethics are considered in the business context (Freeman et al., 2018, 2021). The well-known and highly cited phrase in everyday life "it is just business" is the best example of the resource-based view and its rejection of the social norms with concentration on profit maximization. The citation of the Friedman paper with his "there is one and only one social responsibility of business—to use its resources and engage in activities designed to increase its profits" is the scientific popular analog of the narrowness of the aims of the business from the ethical point of view (Friedman, 2007). By focusing not only on shareholder value maximization but also on the stakeholder requirement inclusion, the stakeholder theory is able to legitimize business in the context of rising global challenges, such as poverty, inequality, and war conflicts that push companies to take action in order to keep the loyalty and well-being of its suppliers, customers, employees, and other key stakeholder groups.

Stakeholder theory also posits that sustainability under stakeholder theory view includes not only financial stability and resilience of the company but also risk management in the environmental and social spheres, thus covering all the triple-bottom line issues (Freeman et al., 2018, 2021). The importance of taking into account the environmental stability is represented by the BP oil spill case—BP's Deepwater Horizon oil spill (2010). It was the largest accidental oil spill in history. The consequences were tremendous. Eleven workers were killed and 17 injured, 134 million gallons of oil were released (according to the findings of the US District Court), and about 2100 km of the US Gulf Coast from Texas to Florida were coated with oil. In the lawsuits that followed, the oil company BP paid $65 billion in compensation to people who relied on the gulf for their livelihoods (Rafferty, 2018).

According to the stakeholder perception of the company, people are viewed as the final end, not as the mean, while resource-based theory suggests people as one of the resources—human resource or labor as a production factor (Freeman et al., 2018, 2021). Person-centered approach in the company is bringing new opportunities in the age of digitalization or fourth industrial revolution that needs new creative tools and approaches, skilled professionals, proper motivation, and common understanding of the values with the employees and counterparties instead of the mechanic interactions with them as resources. Stakeholders, unlike resources, have the ability to influence the company, and thus, the interaction with stakeholders rather than resources is more difficult and more effective at the same time.

Finally, stakeholder theory is mainly about cooperation instead of competition (Freeman et al., 2018, 2021). Resource-based view considered a limited pull of resources with companies battling for the access to such resources. However, in the twenty-first century, some of the most valuable resources are knowledge, reputation, brand, technology, know-how, etc., what can be called intangible resources. They are unlimited by nature and, what is even more important, often need collaboration of the market participants in order to generate such type of the resource. In this context, the need for collaboration appears that may bring benefits to all parties.

To this end, business is starting to change in response to the increasing role of ESG and increasing pressure from stakeholders. First of all, the company's management practice is undergoing changes and, as a result, the entire corporate governance system responsible for the effective integration of ESG aspects into the company's management structures. Since the 1990s, corporations have been under pressure to improve their performance not only economically but also on all the ESG issues. In this context, ESG covers topics related to the environment (e.g., climate change, energy and water use, carbon emissions), social responsibility (e.g., fair trade principles, human rights, product safety, gender equality, health and safety), and corporate governance (e.g., board independence, corruption and bribery, reporting and disclosure, shareholder protection).

2.3 New Perception of Governance Under Stakeholder Theory

According to an early agency theory interpretation of the corporate governance, in particular by Berle and Means (1932), the problem of corporate governance arises from the separation of ownership and control in public corporations and aims at solving the problems of possible opportunistic behavior by managers (agents) toward shareholders (principals). But corporate governance also has to face the problem of the distribution of the quasi rents or returns produced by the firm, which again leads to a conflict between principals and agents and other multiple stakeholders contributing to the generation of quasi rents: employees, customers, suppliers, local community, environment, and society in general. This connects with the question of whose interest the corporation should serve and who should be the beneficiaries of firm-created value.

Definition 2.4 Quasi rents are surplus earnings generated by the factors of production, except land. In other words, abnormal earnings gained by the company.

Broadly speaking, we can differentiate between a shareholder perspective and a stakeholder perspective of the firm (Letza et al., 2004). The shareholding model regards the corporation as a legal instrument for shareholders to maximize their own interests, i.e., investment returns. More in line with the CSR strategy is the stakeholding approach that views the corporation as a locus of responsibility in relation to a wide array of stakeholders' interests or, in words of Tirole (2001, p. 24), as "maximizing the sum of the various stakeholders' surpluses."

According to the stakeholder theory, corporate governance is responsible not only for protecting and increasing shareholder wealth but also for ensuring that strategic decisions are beneficial for all other stakeholders. Such stakeholder-oriented or sustainable corporate governance focuses on the value-creating relationships with all stakeholders, avoiding negative impacts on the environment and society (Lombardi et al., 2019; Sánchez et al., 2011). Moreover, the extension of corporate governance with the firm's key stakeholders is expected to be guided by more than just regulatory or legal requirements to encompass an ethical dimension (Lombardi et al., 2019). Stakeholder orientation creates a competitive advantage for proactive companies both from the possibility of being rewarded by the market and from avoiding risks (Allais et al., 2017).

Governance role is moving from managing the allocation of scarce resources to the management of stakeholders in the value creation process (Elkington, 2006). Since the aim of the governance as well as business itself is changing, the new governance is expected to answer to the new challenges and new questions that arise because of the switch from the resource-centered approach to the stakeholder-oriented. Out of such questions is about the final purpose of the business: whether it is profit generation or broader value generation. What balance should be between the stakeholders and whether shareholders should be still a priority. What is the balance between the environmental, social, and economic goals—should they be taken with the same weight or some of the spheres should be prioritized. What are the tools of achieving the defined above balance and how to measure the effectiveness of the governance under the stakeholder theory perspective. To answer these questions, the concept of sustainable governance is discussed in the following chapters.

References

Allais, R., Roucoules, L., & Reyes, T. (2017). Governance maturity grid: A transition method for integrating sustainability into companies? *Journal of Cleaner Production, 140*, 213–226. https://doi.org/10.1016/j.jclepro.2016.02.069

Aramonte, S., & Zabai, A. (2021). *Sustainable finance: Trends, valuations and exposures*. https://www.bis.org/publ/qtrpdf/r_qt2109v.htm

Berle, A., & Means, G. (1932). The modern corporation and private property in Thomas Clarke. *Corporate Governance: Critical Perspectives on Business and Management*, 173–189.

Brennan, G., & Tennant, M. (2018). Sustainable value and trade-offs: Exploring situational logics and power relations in a UK brewery's malt supply network business model. *Business Strategy and the Environment, 27*(5), 621–630. https://doi.org/10.1002/bse.2067

Business Roundtable. (2022). Retrieved October 7, 2022, from https://www.businessroundtable.org/

Chandler, D. (2016). *Strategic corporate social responsibility: Sustainable value creation* (4th ed.). Sage.

United Nations Development Programme. (2022). *Development challenges and solutions.* Retrieved October 7, 2022, from https://www.undp.org/development-challenges-and-solutions

The European Parliament and the Council of the European Union. (2022). *Directive (EU) 2022/2464 of the European Parliament and of the Council of 14 December 2022 amending Regulation (EU) No 537/2014, Directive 2004/109/EC, Directive 2006/43/EC and Directive 2013/34/EU, as regards corporate sustainability reporting (Text with EEA relevance), EP, CONSIL, 322 OJ L (2022).* Retrieved May 06, 2023, from http://data.europa.eu/eli/dir/2022/2464/oj/eng

Elkington, J. (2006). Governance for sustainability. *Corporate Governance: An International Review, 14*(6), 522–529. https://doi.org/10.1111/j.1467-8683.2006.00527.x

Erhemjamts, O., & Huang, K. (2019). Institutional ownership horizon, corporate social responsibility and shareholder value. *Journal of Business Research, 105*, 61–79. https://doi.org/10.1016/j.jbusres.2019.05.037

European Commission. (2019). *Sustainable corporate governance initiative.* Retrieved October 7, 2022, from https://ec.europa.eu/info/law/better-regulation/have-your-say/initiatives/12548-Sustainable-corporate-governance_en

Freeman, R. E., Dmytriyev, S. D., & Phillips, R. A. (2021). Stakeholder theory and the resource-based view of the firm. *Journal of Management, 47*(7), 1757–1770. https://doi.org/10.1177/0149206321993576

Freeman, R. E., Phillips, R., & Sisodia, R. (2018). Tensions in stakeholder theory. *Business & Society, 1–19*, 213. https://doi.org/10.1177/0007650318773750

Friedman, M. (2007). The social responsibility of business is to increase its profits. In W. C. Zimmerli, M. Holzinger, & K. Richter (Eds.), *Corporate ethics and corporate governance* (pp. 173–178). Springer. https://doi.org/10.1007/978-3-540-70818-6_14

Gautam, D., Goel, R., & Natalucci, F. (2022). Sustainable finance in emerging markets is enjoying rapid growth, but may bring risks. *IMF.* Retrieved October, 7, 2022, from https://www.imf.org/en/Blogs/Articles/2022/03/01/sustainable-finance-in-emerging-markets-is-enjoying-rapid-growth-but-may-bring-risks

IOSCO. (2021). *Environmental, social and governance (ESG) ratings and data products providers.* Final Report Retrieved October 7, 2022, from https://www.iosco.org/library/pubdocs/pdf/IOSCOPD690.pdf

KPMG. (2020). *KPMG international survey of corporate responsibility reporting.* KPMG International.

Letza, S., Sun, X., & Kirkbride, J. (2004). Shareholding versus stakeholding: A critical review of corporate governance. *Corporate Governance: An International Review, 12*(3), 242–262. https://doi.org/10.1111/j.1467-8683.2004.00367.x

Lombardi, R., Trequattrini, R., Cuozzo, B., & Cano-Rubio, M. (2019). Corporate corruption prevention, sustainable governance and legislation: First exploratory evidence from the Italian scenario. *Journal of Cleaner Production, 217,* 666–675. https://doi.org/10.1016/j.jclepro.2019.01.214

Rafferty, J. P. (2018, April 11). 9 of the biggest oil spills in history. *Encyclopedia Britannica.* Retrieved May 6, 2023, from https://www.britannica.com/list/9-of-the-biggest-oil-spills-in-history

Sánchez, J. L. F., Sotorrío, L. L., & Díez, E. B. (2011). The relationship between corporate governance and corporate social behavior: A structural equation model analysis. *Corporate Social Responsibility and Environmental Management, 18*(2), 91–101. https://doi.org/10.1002/csr.244

SASB. (2021). *Standards overview*. Retrieved October 7, 2022, from https://www.sasb.org/standards/

Scatigna, M., Xia, D., Zabai, A., & Zulaica, O. (2021). *Achievements and challenges in ESG markets*. https://www.bis.org/publ/qtrpdf/r_qt2112f.htm

Tantalo, C., & Priem, R. L. (2016). Value creation through stakeholder synergy: Stakeholder synergy. *Strategic Management Journal, 37*(2), 314–329. https://doi.org/10.1002/smj.2337

The European Parliament and the Council of the European Union. (2019). *Regulation (EU) 2019/2088 Of The European Parliament And Of The Council of 27 November 2019*. Retrieved May 05, 2023, from https://eur-lex.europa.eu/eli/reg/2019/2088/oj

Tirole, J. (2001). *Corporate Governance. Econometrica, 69*(1), 1–35. https://doi.org/10.1111/1468-0262.00177

United Nations. (2015). *Sustainable Development Goals*. Retrieved October 7, 2022, from https://sdgs.un.org/goals

Wheeler, D., Colbert, B., & Freeman, R. E. (2003). Focusing on value: Reconciling corporate social responsibility, sustainability and a stakeholder approach in a network world. *Journal of General Management, 28*(3), 1–28. https://doi.org/10.1177/030630700302800301

Chapter 3
Sustainable Governance and Its Components

3.1 Corporate Governance in Detail

Corporate governance is broadly understood as a system by which companies are managed and controlled (The Committee on the Financial Aspects of & Corporate Governance and Gee and Co. Ltd., 1992). Such corporate governance deals with management-shareholder problem and brings profits to the shareholders.

On a practical level, such corporate governance of a company includes several elements that ensure the efficiency of the system.

The board of directors is a main body of the governance system which is responsible for strategizing and oversight and navigation of the company. It defines the company's business priorities and development strategy aimed at profit maximization, increase of the company's assets, and growth of its investment prospects. The board ensures implementation and protection of the rights of the company's shareholders and navigates parties in the conflicts. Minority shareholder rights are at most importance in this case and are to attract special attention. The transparency of an enterprise is as well the board's responsibility: to ensure that company publishes full, fair, and objective information to shareholders and other interested parties. Board establishes internal control system, defines key risk management objectives, and monitors the effectiveness of the system. Monitoring of the results concerns not only internal control function but all functions and company's financial and business operations and the activities of the chief executive body. For being objective in its strategic decisions and control function, the board attracts independent directors that are not related to the company and whose personal well-being is not dependent on the organization decisions. With the aim to separate the strategic and operational tracks, the board is usually represented by the nonexecutive directors, and it is especially relevant for the chief of the board role, where duality of this role with CEO (chief executive officer) may undermine both independence and strategic focus.

A. Cardoni, E. Kiseleva, *Sustainable Governance*, CSR, Sustainability, Ethics & Governance, https://doi.org/10.1007/978-3-031-37492-0_3

Board is supported by its committees, out of which audit committee and nomination and remuneration committee are the most necessary and usually obligatory for public companies. The audit committee performs a preliminary review and makes recommendations to the board of directors on issues related to the control of the company's financial and business operations. The main task of the nomination and remuneration committee is to develop and submit recommendations to the company's board on remuneration policies, assessment of the efficiency of the management bodies, continuous monitoring of compliance of the criteria, and remuneration policy with the corporate strategy, financial position, etc.

The policies and procedures as another key element of the governance system ensure the standardized approach to all similar situations and verify that actions are sound and correspond to the law and main aim of profit maximization. The internal control and audit verify that the policies are implemented and there are no cases of wealth maximization for separate persons instead of a company in general.

Finally, there is an external element of the system—law of the country and institutions such as authorities, central banks, or international organizations. Such external institutions provide the obligatory or recommended norms for companies to follow in building their governing bodies. The norms are usually applied to public and large corporations, while small and medium enterprises use the norms as an example of best practice.

3.2 Sustainable Corporate Governance

Continuing the discussion of Chaps. 1 and 2, sustainable corporate governance is the next stage in the development of the currently accepted understanding of corporate governance based on the theory of agency relations (Eisenhardt, 1989). In contrast to the classical interpretation of corporate governance, sustainable corporate governance is based on stakeholder theory (Freeman, 1984), significantly expanding the list of participants in corporate relations and expanding the number of elements of corporate governance.

The sustainable governance may be called different terms, such as stakeholder governance, multi-actor governance, ethical governance, or others.

In the context of society at large, governance may be defined as "the sum of the formal and informal rule systems and actor-networks at all levels of human society that are set in order to influence the co-evolution of human and natural systems in a way that secures the sustainable development of human society" (Biermann, 2007).

Going back to the corporate level and adjusting the abovementioned logic, sustainable governance is organized so that the company serves to sustainable development of the society. In this sense, sustainable corporate governance is a contribution that corporations make to the environmental safety, social stability, cultural and ethical behavior, and the creation of activities that bring outcomes, profits, and self-fulfillment to those undertaking it.

To this end, the beneficiaries of the corporate operations enlarge and include not only shareholders but also representatives of other interested parties that are called stakeholders. According to the most cited author of stakeholder theory, R. Edward Freeman, who originally detailed the stakeholder theory out of organizational management and business ethics, stakeholders are "those groups without whose support the organization would cease to exist" (Freeman, 1984): customers, employees, creditors, investors, suppliers, contractors. In addition to ones who are making an impact on the organization, there are some stakeholders who are dependent on the organization's resources and operations or are affected by the social or environmental consequences of company's operation. Such stakeholders include local communities, society at large, and sometimes environment that is considered as a nonhuman stakeholder.

Definition 3.1 Stakeholder is a person or an organization that "can reasonably be expected to be significantly affected" by company's activities (GRI, 2023).

When company generates value together with stakeholders and for stakeholders, such value is usually called a cocreated value or sustainable value. The clearest definition of sustainable value is proposed by Hart and Milstein (2003) in the much cited paper "Creating Sustainable Value." Hart and Milstein (2003) defined sustainable value as "strategies and practices that contribute to a more sustainable world while simultaneously driving shareholder value" (Hart & Milstein, 2003). The majority of authors put an emphasis on creating sustainable value for a wide range of stakeholders, including those who have no power to influence the company but are affected by the company's activities. Outside stakeholders, such as suppliers, customers, employees, or even environment as a nonhuman stakeholder, are keen for sustainable value creation and are the main characteristic of sustainable value. In an interconnected and interdependent system, each stakeholder must be a means and an end. Each contributes to collective benefit and each must also receive the part of the overall wealth (Freeman et al., 2018). Business integrates stakeholders as far as drivers push toward such integration: stakeholder tension is becoming stronger, while increase in the population, economic fluctuations, and recent COVID-19 pandemic lead to the cross-dependence between the company and the well-being of local economics and communities (Cardoni et al., 2020b).

Keeping in mind the prevalence of the profit maximization for corporations, the overall definition of sustainable corporate governance may be as a corporate governance system that is based on the integration of the goals of shareholders and key stakeholders, aimed at value maximization, protecting the interests of society and the environment as a whole (Allais et al., 2017; Cardoni et al., 2020a; Lombardi et al., 2019).

In practice, the company's sustainable corporate governance is aimed both at the productive resolution of the conflict of interests between the shareholder and management and at protecting the rights and interests of key stakeholders. This is possible through fair and public coordination of the company's for-profit goals and stakeholder requests by creating opportunities for stakeholders to participate in the management of the company in order to legitimate the business operation. Company uses complex management mechanisms in achieving this aim:

- Development of ESG strategies and ESG risk management.
- Standardization of various types of operational management (quality management, occupational safety and industrial safety, environmental management, energy management, carbon management, etc.)
- Inclusion of stakeholders in the committees of the board of directors and specialized public councils under the executive director.
- Introduction of anti-corruption policies and systems of relevant measures.
- Development of practices to ensure transparency of the company and compliance with business ethics rules and norms.

The notion of sustainable corporate governance is gaining popularity not only among theoretical works but also among policymakers. In July 2020, the European Commission published the "Study on Directors' Duties and Sustainable Corporate Governance." The report aimed at finding an evidence for the need to stop short-termism in EU corporate governance and recommended many changes to support sustainable corporate governance. In February 2022, after a period of the consultations with key stakeholders, the Commission issued a Proposal for a Directive on Corporate Sustainability Due Diligence Directive amending document of 2019. According to the Directive, large EU organizations, EU companies from high-impact sectors, and large non-EU companies with tight business interconnections with the Union will be required to identify, prevent, mitigate, and account for their "adverse impacts" on human rights and the environment. Adverse impacts are defined by reference to violations of prohibitions, obligations, and rights enshrined in the international conventions listed in the Annex to the Proposed Directive (European Commission, 2019).

3.3 Discussions over Components of Sustainable Governance

The definition of sustainable (ethical) corporate governance is new and differs across authors, but it has several common characteristics repeating in different works. All authors agree that sustainable governance is about the following:

- Combination of the goals of shareholders with the interests of a wide range of stakeholders.
- Cocreation of the value together with stakeholders and for stakeholders instead of profit maximization.
- Both short- and long-term orientation instead of the short-termism.

However, opinions differ as to what is included in such ethical governance and what are the metrics that allow assessing the level of sustainability of corporate governance in the company.

According to Klettner et al. (2014), sustainable governance is about corporate policies and procedures that ensure the company's commitment to the implementation of environmentally and socially significant initiatives, including the following:

- Communication tools (reporting in the field of sustainable development, interaction with stakeholders).
- Tools for demonstrating commitment to the principles of sustainable development (voluntary participation in thematic ESG initiatives).
- Management mechanisms (composition of the board of directors, composition of senior management).
- Tools for implementing sustainable development practices (i.e., remuneration system).

According to Allais et al. (2017), corporate governance includes the following:

- Consultations with stakeholders and their participation in decision-making.
- Sustainable and mutually beneficial relationships with local participants.
- The strategy of creating intangible assets from the point of view of a set of values (through external networks and partnerships).
- Activities aimed at reducing the impact of business on the environment.
- Activities aimed at creating value to meet the needs and expectations of primary and secondary stakeholders (Allais et al., 2017).

Shiroyama et al. describe sustainable governance with a wider look at the concept and compare the sustainable governance with the Japanese notion of "doushouimu" that literally means "sharing the same bed, dreaming different dreams" (Shiroyama et al., 2012). In this case, sustainable corporate governance, or multi-actor governance, includes three elements:

- Mechanisms that enable realization of different stakeholder requests through the organization.
- Mechanisms of innovation that produce benefits to be allocated among actors and different spatial levels of governments.
- Transparency and fairness of rules of allocation.

On the contrary, Kolk (2008) interpretation is more practical and discusses corporate governance structure in terms of governance bodies (Kolk, 2008):

- External or internal advisory councils (on environmental issues, social responsibility) and special committees under the board of directors.
- Department responsible for ESG direction and accountable to the CEO.
- Sustainability experts in business units.
- Compliance mechanisms related to sustainability (environmental and social policies, code of ethics, hotline, etc.)

Xie et al. made an attempt to operationalize the criteria of sustainable corporate governance in the organization. Among such practically oriented criteria (Xie et al., 2019) are as follows:

- GRI reporting.
- Audit of nonfinancial reporting.
- Signature of UN Global Compact initiatives.
- Independence of directors.
- Meetings of the audit committee.
- Presence of women on the board.
- ESG KPI.
- Code of ethics.

According to the EU Corporate Sustainability Due Diligence Directive, the main components are assumed to be related to compliance and due diligence in the field of sustainability. The company is supposed to have the corporate policy in the field of due diligence including a code of conduct for employees and subsidiaries. The measures and actions from the company are expected to identify, prevent, mitigate, or bring to an end the adverse impacts from its operations. Out of the measures, there may be as follows:

- Development and implementation of preventive and/or corrective action plans.
- Contractual assurances from the business partners to comply with the code of conduct.
- Investments into management or production processes and infrastructures.
- Support to SMEs.
- Financial compensation to affected persons and communities if adverse impacts have happened.
- Termination of the business relationships with non-reliable partners.

Company should organize a safe process for stakeholders to submit complaints regarding the company's adverse impacts. The monitoring of the effectiveness of the measures undertaken should be in place. After all, the information about the compliance measures should be publicly disclosed (European Commission, 2019).

3.4 Common Blocks of Sustainable Corporate Governance

Despite the heterogeneity in the literature, there are several components representing the sustainable governance that appear in the majority of the scientific papers and in the frameworks developed by the NGOs in the field.

A great deal in defining the most accurate metrics for governance in the context of ESG is undertaken by the World Economic Forum, with its White Paper "Measuring Stakeholder Capitalism" (World Economic Forum, 2020). This document was developed in collaboration with Deloitte, EY, KPMG, and PwC and went through the consultations with more than 200 companies, investors, and other key players.

The WEF 2020 White Paper identifies a set of universal, material ESG metrics and recommended disclosures that could be found in the mainstream annual or sustainability reports of companies from different industries and countries. The main role of such indicators was to demonstrate companies' commitment to long-term sustainable value creation. The description of the common theoretical blocks of sustainable governance is largely based on the WEF 2020 White Paper due to its popularity among both academic researchers and practitioners (Table 3.1).

Each of the abovementioned blocks of the corporate governance is stated to be vital for value creation process. In all the cases, the value generation is presented by the increase of the effectiveness of ESG initiatives, the risk mitigation, brand development, and increase of stakeholder loyalty.

3.5 The Role of Sustainable Governance Blocks in Value Generation

3.5.1 ESG Strategy and Risk Management

The strategic response to stakeholders' inquiries is accepted to be vital for gaining competitive advantage (Ioannou & Serafeim, 2019; Porter & Kramer, 2006). The inclusion of ESG factors into investors' decisions (responsible investments) started as a faith-based rejection of some unethical company's activities, while today it is more oriented at gaining financial benefits from the strategic incorporation of particular E, S, and G issues (Orlitzky, 2015).

This thesis was explained by Porter and Kramer (2006) on a theoretical basis. "For any company, strategy must go beyond best practices. It is about choosing a unique position—doing things differently from competitors in a way that lowers costs or better serves a particular set of customer needs." Authors state that strategic approach moves way beyond good corporate citizenship, mitigation of negative impacts, or small ad hoc social and environmental projects. Strategic ESG involves incorporation of its principals at its core. While creating a business model and planning the product outlook, company should consider both inside-out and outside-in dimensions working in tandem to benefit both society and a company's own competitiveness. In this case, company may use ESG for value generation rather than irrelevant cost creation. Authors also provide an example of Toyota's Prius, the hybrid electric/gasoline vehicle, the first in a series of innovative car models that have produced competitive advantage and environmental benefits by answering the carbon emissions concerns and aim of profit maximization for car manufacturing firm (Porter & Kramer, 2006).

On a practical level, the same thesis about importance of strategic approach is supported at Ioannou and Serafeim (2019) research on 2095 firms for the period 2012–2019. In their paper, they explored the adoption of sustainability actions and analyzed a fundamental issue for ESG field, finding the conditions under which

Table 3.1 Common blocks of sustainable corporate governance

Section	Description
ESG strategy and risk management	Companies have a stated goal related to economic, social, and environmental performance in the overall business strategy or in a separate ESG strategy; the risks from the economic, social, and environmental spheres; as well as the relevant risks of the modern society are stated, such as the risks of the fourth industrial revolution (data security/ digitalization, etc.), pandemic risks, and the risks of climate change. Companies disclose information about the impact of those risks on their business and indicate risk appetites, dynamics of changes in risk levels, and the company's response to identified risks
Board composition and stakeholder engagement	This block describes such qualitative characteristics of the board of directors: • Presence or absence of a combination of the positions of chairman of the Board of Directors and Chairman of the management board. • Independence of the majority (≥51%) of the members of the board of directors. • Presence of women on the board. • Duration of the directors' tenure (should not be too short or too long). • Previous significant positions and obligations of each of the directors (whether it provides them with relevant experience and skills for dealing with ESG risks and opportunities). Together with sustainable board structure, company may engage stakeholders in the decision-making process by including stakeholders in the process of identifying significant topics for disclosure in the annual reports and by creating a stakeholder/ESG or similar committee under the board of directors
Anti-corruption and business ethics	Companies develop and approve anti-corruption policies, norms, and standards, develop tools for reporting unethical or illegal behavior (e.g., a hotline), discuss anti-corruption initiatives, and interact with stakeholders through conferences and workshops and public initiatives and who are signatories of the public anti-corruption initiatives. Employees of such companies are trained in anti-corruption policy and anti-corruption procedures, members of governing bodies are also familiarized with modern trends in the field of anti-corruption, and an anti-corruption clause is prescribed in contracts for suppliers and contractors. Companies are transparent in the field of anti-corruption: They disclose information about the number of employees and managers who have completed anti-corruption training programs, as well as the number of corruption cases that occurred in the reporting period
Transparency	Company provides public information on the corporate governance elements on a timely and regular basis. The most reliable source of information for stakeholders is a report signed by the CEO of the company. Corporate governance information may be presented in the usual annual report or sustainability report, while the latter can be called ESG report/impact report/report

(continued)

Table 3.1 (continued)

Section	Description
	on sustainable development of the company/integrated report or other synonyms. The website information and information on corporate social media sources may be also of a great importance for day-to-day communication with key stakeholders, especially during the periods of rapid changes in the company's operations

Source: constructed by the author

companies are able to maintain their sustainability-based differentiation advantage. The authors studied the topic on the basis of the companies with high-level ESG practices operating in developed economies; thus, the sample of the researchers was not only under pressure from stakeholders to legitimate company's activities but also under imitation pressures by industry peers. The results over more than 2000 companies show that even though sustainability actions are increasingly imitated within industries, some companies can maintain their competitive advantage by undertaking unique strategic actions, characterized by a high degree of novelty, thus protecting the valuable industry positions that they occupy (Ioannou & Serafeim, 2019).

3.5.2 *Board Composition and Stakeholder Engagement*

The board of directors is the most important governance mechanism within the company, and therefore its composition, in terms of gender, age, nationality, and professionalism, of the components is considered a crucial determinant of the performance of the organization in general and of sustainability performance, in particular (Naciti et al., 2022).

There are plenty of empirical studies that support such evidence. The study of Konadu et al. (2021) explored corporate governance pillars comprising board functions, structure, strategy, compensation, and shareholder rights with the database of listed S&P 500 firms. According to the authors, business sustainability in environmental, social, and financial dimensions is impacted positively by board functions and board structure. At the same time, mismanagement of stakeholder communication leads to poor results in all triple-bottom line performance (Konadu et al., 2021).

There are several elements of corporate structure that appear in the companies and represent the best practice examples in building sustainable corporate governance.

Starting with the board of directors, being the main body, it plays an important role in building an effective ESG strategy. In addition to the board, there may be a separate ESG board committee that is specialized in the fields of environmental, social, and stakeholder communication fields. Such a body approves the goals and strategy of sustainable development, sets priorities, and monitors activities.

The allocation of such a body makes it possible to build a regular system of relationships with stakeholders: to hold meetings, round tables, conferences, and other events in which interested parties evaluate the company's activities in terms of their requests and make recommendations for the future.

Example 3.1 Nestlé Creating Shared Value Council
Nestlé presents practice of completely external stakeholder committee. Apart from the ESG and Sustainability Council (includes Executive Board Members) and Sustainability Committee under the Board, the company has the Creating Shared Value Council that is an external advisory council.

The Creating Shared Value (CSV) Council helps ensure the sound development of long-term sustainability and positive social and economic impacts of the CSV business strategy. The Nestlé CSV Council comprises nine external members, whose expertise spans corporate social responsibility, strategy, sustainability, nutrition, water, and rural development. It is made up of global leaders in the fields of nutrition, water, rural development, business strategy, and corporate responsibility. The members serve as advisors to help facilitate the ongoing development of the company's Creating Shared Value business strategy.

Going down in the corporate hierarchy, companies may have an ESG executive body (e.g., a sustainability department). It directly implements projects in the field of sustainability and is usually subordinate to the CEO of the company. Its functions include participation in the development of a sustainable development strategy, managing its implementation, and adjusting the strategy depending on external challenges. Such department also monitors the implementation of internal policies in the field of sustainable development (these include policies in the field of ethics, ecology, personnel management, etc.) and prepares annual reports: both internal for the Board of Directors and public nonfinancial reporting for a wide range of stakeholders. The executive body also controls and regulates subordinate structural units in the field of sustainable development or navigates other departments or functional units of the company in mutual sustainability projects.

In the process of the navigation of the different departments under one roof of sustainability, companies may have a coordination body for functions/divisions and subsidiaries. Such a body is collegial: representatives meet once a month, a quarter, or a year. The task of the coordination body is to bring together representatives from different functional areas and divisions, as well as from subsidiaries of the company in meetings for the joint development of sustainability projects.

Research centers on ESG topics are often created as a separate department or as a structural unit independent of the company, which includes external experts, representatives of research centers, and universities. Usually, such centers are specialized on a particular issue of sustainable development, like malnutrition or battling pandemic: they develop and conduct research and offer specific recommendations on certain aspects of ESG.

Charitable foundation is another element that helps to structure the company's social and environmental initiatives and ensure transparency in the financing of such initiatives. The task of the fund is to achieve maximum efficiency of social investments. It allows attracting additional funds for social investments and at the same time increases the transparency of charitable activities. The foundation develops a social investment strategy, plans long-term charitable programs, accumulates the necessary funds, forms a portfolio of projects, and implements them. The foundation may be created to solve a specific social problem or to manage the entire charitable activities of the company.

3.5.3 Anti-corruption and Business Ethics

Anti-corruption is a hot topic of nowadays that requires rigorous attention. Unlike other blocks of corporate governance, this block is about mitigating risks rather than creation of value.

The annual cost of corruption is estimated to be equal to US$1.5–2 trillion a year (Kaufmann, 2005). This amount accounts to roughly 2% of 2017 global GDP (World Bank, 2018). The United Nations included anti-corruption in the global agenda with the help of Sustainable Development Goals (SDGs). SDGs stimulate action in areas of critical importance for humanity and the planet and declare the goal 16 "Pease, Justice and Strong Institutions" with the target on substantial reduction of the corruption and bribery in all their forms (United Nations, 2015).

The topic of anti-corruption concerns not only country-level issue but also company-level agenda. Almost one in five firms worldwide reports receiving at least one bribery payment request when engaged in regulatory or utility transactions (UN Global Compact, 2019). Such statistics undermines the confidence and trust in business among various stakeholders: investors, customers, employees, and the public. There have been a lot of business ethics scandals worldwide showing the dramatic fall of trust. For example, one of the most famous scandals, the Enron scandal of 2001, had far-reaching political and financial implications. The Enron Corporation, an energy company and America's seventh largest company employing 21,000 staff in more than 40 countries, went bankrupt after the incident of accounting corruption issue (BBC, 2002). Despite the well-known consequences of corruption, the episodes still appear.

Example 3.2 Hotline as the most popular anti-corruption tool
In order to mitigate the misconduct and find the cases of misbehavior, companies motivate employees to report wrongdoing in an organization through

(continued)

Example 3.2 (continued)
safe and anonymous ways of communication. With this aim, the best approach is to ensure the availability of online tool that is managed by the third party, where employee may apply its concerns without a fear of pressure from the colleagues or counterparties.

For example, Continental Gold, a Chinese mining company, is using a third-party payment whistleblowing platform called línea transparencia. Whistleblowers using this platform do not need to have direct contact with Continental Gold. Instead, they report through an encrypted third-party whistleblowing channel (telephone or email). After the platform collates and processes the information, it forwards the information to the person from Continental Gold in charge of receiving such reports. As the information is professionally processed by the intermediary platform, its confidentiality is greatly increased, eliminating the worries of the whistleblowers.

The whistleblowing system is maintained and managed by the company's Supervision and Audit Office, with dedicated personnel responsible for receiving grievance information. The whistleblowing methods are published on the company's official website are included in the clauses of all external contracts.

The whistleblowing process consists of several steps. After receiving the information, company analyzes the content and categorizes such information for further action. Reports on serious issues are investigated and handled by company or even reported to the police (Zijin Mining Group Co., Ltd., 2022).

3.5.4 Transparency

According to stakeholder theory, the disclosure of nonfinancial information is part of the dialog between a company and its stakeholders, and it provides information on a company's activities that legitimize its behavior, educate and inform, and change perceptions and expectations. What is more, company may find the ways for generating competitive advantage out of the long-term and mutually profitable relations with key stakeholders. To this end, stakeholder engagement is essential to develop an understanding of stakeholders' expectations and co-opt important external requests.

In agency theory perspective, corporate disclosure policies are aimed at legitimizing the company's activities to stakeholders, given their diverse and various expectations. Companies manage their legitimacy by signaling to stakeholders that their behavior is appropriate and desirable.

Good corporate governance and sustainability disclosure can be seen as complementary mechanisms used by companies to enhance relations with stakeholders.

There is a variety of the nonfinancial reports and even more nonfinancial standards that may be used by a company. However, there are few most widely accepted.

Nonfinancial reports include information on sustainable corporate governance in a most overwhelming and reliable way. There are several types of reporting frameworks that are used by researchers who use nonfinancial reports in their studies. One of the most popular is GRI—Global Reporting Initiative. Global Reporting Initiative (GRI) guidelines, developed in 1997, deserve a particular attention as they are today the most widely used. The standards are designed to enhance the global comparability and quality of information on ESG impacts, thereby enabling greater transparency and accountability of organizations (GRI, 2023). With a modular, interrelated structure, they represent the global best practice for reporting on a range of environmental, economic, and social impacts.

Integrated Reporting (<IR>) framework is also one of the most widely used by researchers. <IR> applies principles and concepts that are aimed at making the reporting process more close-knit and efficient in order to improve the quality of information available to investors. In 2022, the Integrated Reporting framework became a part of the IFRS Foundation, the standard setter of the financial reporting. This event may lead to a great collaboration and boost in the standard popularity.

The International Financial Standards Board has recently postulated its intention to bring the harmony into the standards heterogeneity. On March 2022, the Exposure Draft IFRS Sustainability Disclosure Standard was issued. The document focuses on the global coherence of the nonfinancial reporting standards, most pressingly on climate-related disclosures, focusing on the capital market participants as main stakeholders of the sustainability reporting. Such way of focus on one block of stakeholders limits the possibilities to answer the various stakeholder inquiries, but it brings homogeneity in the part of ESG data market boosting the development of unified approach to ESG information aggregation and analysis (IFRS, 2021).

References

Allais, R., Roucoules, L., & Reyes, T. (2017). Governance maturity grid: A transition method for integrating sustainability into companies. *Journal of Cleaner Production, 140*, 213–226. https://doi.org/10.1016/j.jclepro.2016.02.069

Biermann, F. (2007). 'Earth system governance' as a crosscutting theme of global change research. *Global Environmental Change, 17*(3), 326–337. https://doi.org/10.1016/j.gloenvcha.2006.11.010

Cardoni, A., Kiseleva, E., & Lombardi, R. (2020a). A sustainable governance model to prevent corporate corruption: Integrating anticorruption practices, corporate strategy and business processes. *Business Strategy and the Environment, 29*(3), 1173–1185. https://doi.org/10.1002/bse.2424

Cardoni, A., Kiseleva, E., & Taticchi, P. (2020b). In search of sustainable value: A structured literature review. *Sustainability, 12*(2), 2. https://doi.org/10.3390/su12020615

Eisenhardt, K. M. (1989). Agency theory: An assessment and review. *Academy of Management Review, 14*(1), 57–74. https://doi.org/10.5465/amr.1989.4279003

BBC. (2002). *Enron scandal at-a-glance*. Retrieved 7 August 2022, from http://news.bbc.co.uk/2/hi/business/1780075.stm

European Commission. (2019). Proposal for a Directive of The European Parliament and of The Council on Corporate Sustainability Due Diligence and amending Directive (EU) 2019/1937.

Retrieved August 7, 2022, from https://eur-lex.europa.eu/legal-content/EN/TXT/?uri=CELEX%3A52022PC0071

Freeman, R. (1984). *Strategic management: A stakeholder approach*. Cambridge University Press.

Freeman, R. E., Phillips, R., & Sisodia, R. (2018). Tensions in stakeholder theory. *Business & Society, 1–19*, 213. https://doi.org/10.1177/0007650318773750

GRI. (2023). *Consolidated set of the GRI standards*. Retrieved 06 May 2023, from https://www.globalreporting.org/how-to-use-the-gri-standards/gri-standards-english-language/

Hart, S. L., & Milstein, M. B. (2003). Creating sustainable value. *Academy of Management Perspectives, 17*(2), 56–67. https://doi.org/10.5465/ame.2003.10025194

IFRS. (2021). *Exposure Draft and comment letters: General Sustainability-related Disclosures*. Retrieved August 7, 2022, from https://www.ifrs.org/projects/work-plan/general-sustainability-related-disclosures/exposure-draft-and-comment-letters/

Ioannou, I., & Serafeim, G. (2019). Corporate sustainability: A strategy? (SSRN Scholarly Paper No. 3312191). https://doi.org/10.2139/ssrn.3312191

Kaufmann, D. (2005). *Myths and realities of governance and corruption* (SSRN scholarly paper ID 829244). Social Science Research Network. https://papers.ssrn.com/abstract=829244

Klettner, A., Clarke, T., & Boersma, M. (2014). The governance of corporate sustainability: Empirical insights into the development, leadership and implementation of responsible business strategy. *Journal of Business Ethics, 122*(1), 145–165. https://doi.org/10.1007/s10551-013-1750-y

Kolk, A. (2008). Sustainability, accountability and corporate governance: Exploring multinationals' reporting practices. *Business Strategy and the Environment, 17*(1), 1–15. https://doi.org/10.1002/bse.511

Konadu, R., Ahinful, G. S., & Owusu-Agyei, S. (2021). Corporate governance pillars and business sustainability: Does stakeholder engagement matter? *International Journal of Disclosure and Governance, 18*(3), 269–289. https://doi.org/10.1057/s41310-021-00115-3

Lombardi, R., Trequattrini, R., Cuozzo, B., & Cano-Rubio, M. (2019). Corporate corruption prevention, sustainable governance and legislation: First exploratory evidence from the Italian scenario. *Journal of Cleaner Production, 217*, 666–675. https://doi.org/10.1016/j.jclepro.2019.01.214

Naciti, V., Cesaroni, F., & Pulejo, L. (2022). Corporate governance and sustainability: A review of the existing literature. *Journal of Management and Governance, 26*(1), 55–74. https://doi.org/10.1007/s10997-020-09554-6

Orlitzky, M. (2015). The politics of corporate social responsibility or: Why Milton Friedman has been right all along. *Annals in Social Responsibility, 1*(1), 5–29. https://doi.org/10.1108/ASR-06-2015-0004

Porter, M. E., & Kramer, M. R. (2006). Strategy and society: The link between competitive advantage and corporate social responsibility. *Strategic Direction, 23*(5). https://doi.org/10.1108/sd.2007.05623ead.006

Shiroyama, H., Yarime, M., Matsuo, M., Schroeder, H., Scholz, R., & Ulrich, A. E. (2012). Governance for sustainability: Knowledge integration and multi-actor dimensions in risk management. *Sustainability Science, 7*(1), 45–55. https://doi.org/10.1007/s11625-011-0155-z

Shumanov, I. (2019, March 7). Troika Bank Scandal Exposed Global Failure to Prevent Money Laundering. *The Moscow Times*. https://www.themoscowtimes.com/2019/03/07/troika-bank-scandal-exposed-global-failure-to-prevent-money-laundering-a64730

The Committee on the Financial Aspects of & Corporate Governance and Gee and Co. Ltd. (1992). *The Financial Aspects of Corporate Governance* (ISBN 0 85258 913 1). Retrieved August 7, 2022, from https://www.icaew.com/-/media/corporate/files/library/subjects/corporate-governance/financial-aspects-of-corporate-governance.ashx?la=en

UN Global Compact. (2019). *The Sustainable Development Goals Report 2018*. Retrieved August 7, 2022, from https://unstats.un.org/sdgs/files/report/2018/TheSustainableDevelopmentGoalsReport2018-EN.pdf

United Nations. (2015). *Transforming Our World: The 2030 Agenda for Sustainable Development* (Resolution Adopted by the General Assembly on 25 September 2015). Retrieved August 8, 2022, from https://doi.org/10.1891/9780826190123.ap02

World Bank. (2018). *World Bank national accounts data*. Retrieved September 2, 2022, from https://data.worldbank.org/

World Economic Forum. (2020). *Measuring Stakeholder Capitalism: Towards Common Metrics and Consistent Reporting of Sustainable Value Creation*. Retrieved September 2, 2022, from https://www.weforum.org/reports/measuring-stakeholder-capitalism-towards-common-metrics-and-consistent-reporting-of-sustainable-value-creation/

Xie, J., Nozawa, W., Yagi, M., Fujii, H., & Managi, S. (2019). Do environmental, social, and governance activities improve corporate financial performance? *Business Strategy and the Environment, 28*(2), 286–300. https://doi.org/10.1002/bse.2224

Zijin Mining Group Co., Ltd. (2022). *2021 environmental, social and governance report*. http://www.zijinmining.com/upload/file/2022/05/06/19456a9811b349408f8c4365972740a9.pdf

Chapter 4
Measurement of Sustainable Governance

4.1 Academic Approach to Measuring Sustainable Governance

Academic research in the field of ESG shifted from general and broad research questions (such as the impact of firms on society) to more specific aspects related to the mechanisms that explain and drive firms' sustainable behavior (such as the role played by the composition of the board of directors). For example, researchers changed the set of keywords used in their papers from "Society," "Innovation," and "Commitment" to "Social responsibility," "Sustainability report," "Director," and "Board" to "Board size," "Independent director," and "Female director" (Naciti et al., 2022).

The same change is visible in relation to the sustainable corporate governance research that goes from the description of the general frameworks to operationalization of the concept, even though the operationalized metrics vary greatly due to the novelty of such stream of literature.

The majority of authors focus on the *board of directors' characteristics* in attempt to evaluate whether governance of the company is able to contribute to sustainable development of the organization. For example, Lu (2021) explores four board attributes: board size, board independence, CEO duality (where the CEO is also the chairperson of the board), and female directors on board (Lu, 2021). The same criteria are used by Peng and Zhang (2022) in their recent research on the relationship between corporate governance and environmental sustainability performance. Many more other authors state similar criteria while measuring sustainable development of the corporate governance system.

Out of the board characteristics, the most used is *independency of directors* (Sánchez et al., 2011; Hussain et al., 2018; Lu, 2021; Peng & Zhang, 2022). Independent directors are generally under less pressure from managers and shareholders and more sensible to increased awareness of corporate reputation risks and various stakeholder group requests. Independent directors tend to promote higher

A. Cardoni, E. Kiseleva, *Sustainable Governance*, CSR, Sustainability, Ethics & Governance, https://doi.org/10.1007/978-3-031-37492-0_4

requirements for corporate transparency, effectively reducing the possibility of information asymmetry. *Board size* is another metric used by researchers, which is stated to be important for strategic decision-making and management supervision: large board can promote and strengthen control function, while different members with vast backgrounds are able to improve board knowledge and experience, reduce information asymmetry, lessen potential managerial opportunism, and improve board decision-making on ESG issues. Regarding *CEO duality* metric, it is suggested that management and supervision should be separated to reduce CEOs' opportunism and enhance board attention toward environmental responsibility in decision-making. Several studies find that the inclusion of *women on boards* is beneficial for companies as it helps mitigate risks and encourages prudent financial decisions (Sánchez et al., 2011). According to Dicuonzo (2022), the figure of the *sustainability manager* helps to unite the profit-maximization objectives with the impact on the community, the environment, and all stakeholders, overseeing the social responsibility of the company (Dicuonzo et al., 2022). *Sustainability committee* is often referred to as best practice criteria when measuring the corporate governance effectiveness for sustainable development of the organization. Such committee assists the board of directors in supervising and evaluating the company's responsibility practices, evaluates dangerous ESG risks, and allows to take stakeholder's opinions onboard (Hussain et al., 2018).

Risk management is a less popular but not less important area of investigation in relation to sustainable governance (Allegrini & D'Onza, 2011). Dicuonzo (2022) in his research observes whether the main ethical risks, such as reputational risk, misconduct risk, sustainability risk, and crisis management systems, are observed by the company. The examination of the new emerging risks, such as climate change risk, is considered an essential tool, especially after the implementation of the Paris Agreement, in which it prioritizes investor awareness of the risks and opportunities associated with climate change (Dicuonzo et al., 2022).

Compliance metrics may refer to the presence or absence of an external auditor of the nonfinancial report, remuneration practices. Remuneration systems are supposed to take into account an extra-financial performance-oriented compensation policy, nonfinancial performance remuneration targets, and the clawback or malus clauses aimed at the recovery of the variable compensation in the presence of unethical behavior by managers (Dicuonzo et al., 2022).

Finally, authors evaluate how companies *communicate* their sustainability activities externally with the following important documents: the integrated annual report, annual report, nonfinancial report, or alike. Some focus also on the adoption of the Global Reporting Initiative (GRI) guidelines, Integrated Reporting framework, TCFD (Task Force on Climate-Related Financial Disclosures), or other popular guidance on corporate ESG reporting (Dicuonzo et al., 2022; Jacoby et al., 2019).

External elements of corporate governance are also included into the measurements by some authors (Asif et al., 2011; Dicuonzo et al., 2022).

External norms are usually presented by the *law and regulations* imposed as an obligation or recommendation for companies in each country as well as policies, certifications, and awards given to the company by the third parties. Out of them, one of the most popular is ISO 14000 or EMS (environmental management system) certification, the Equator Principles (EP), and a code of conduct. Specifically, some companies choose to be certified to recognized international standards, such as ISO 14001, which is an international, voluntary certification scheme for an environmental management system administered by the International Organization for Standardization (ISO). Indeed, the implementation of such policies motivates proper monitoring of the ESG activities of the company and its environmental and social impacts, helps in formulating corporate ESG purposes, and communicates with stakeholders during the strategy development and implementation processes.

Overview 4.1 EU regulation on corporate governance

As one of the most rapidly changing corporate governance environments in the world, Europe represents a highly developed system of regulations in the corporate governance field, leading the move to the stakeholder-centered governance.

Back in 2006, European Commission Directive 2006/46/EC required all listed companies to produce a corporate governance statement in their annual report to shareholders. By this, EU regulation covers at least partially two blocks of sustainable corporate governance: transparency and board structure.

Anticorruption block of EU companies is highly regulated as well with the help of such regulative documents as the following:

- The 1997 Convention on fighting corruption involving officials of the EU or officials of EU countries.
- The 2003 Council Framework Decision on combating corruption in the private sector, which criminalizes both active and passive bribery.
- The 2008 Council Decision 2008/852/JHA.

EU made a step into the development of the ESG risk management block of sustainable corporate governance. The proposal for a Directive on Corporate Sustainability Due Diligence was adopted by the Commission on February 23, 2022, aiming at better enabling companies to identify and mitigate actual or potential human rights and environmental adverse impact in the companies' operation and value chains (European Commission, 2022).

The role of stakeholders has not been regulated yet at EU level and varies considerably across companies, sectors, and countries. The rights of stakeholders, such as codetermination and worker protection, are protected by company law or other relevant legislation in some European nations. Companies in other nations, on the other hand, typically concentrate more narrowly on the interests of shareholders.

Some authors also included the external factor of *national culture*. For example, Peng and Zhang (2022) used it as a moderating variable in the equation, where authors measure masculinity and uncertainty avoidance as related metrics (Peng &

Zhang, 2022). In the context of emerging markets, cultural aspect becomes even more important.

Example 4.1 National culture impact on corporate governance: Russian case

Mccarthy and Puffer (2008) used the integrative social contracts theory to adopt the agency theory to the local Russian market to describe local ethical norms.

Authors highlighted the importance of personal networks in Russian business. While having a list of positive impacts of such approach to governance, there are several weak points that impact negatively on financial results of the companies. In particular, these are three dimensions: trust, disrespect for private property, and personal favors. According to Mccarthy and Puffer (2008), these traditional, culturally based local norms are likely to prevail for some time in the decision-making process of the part of Russian managers and board directors (Mccarthy & Puffer, 2008).

Thus, the main role of the sustainable corporate governance in Russia is twofold. Firstly, it needs to align local norms (trust, disrespect for private property, and personal favors) to internationally accepted hyper-norm of law prevalence. Sustainable governance also should broaden the in-group wealth maximization orientation to the more standardized perspective, where the company itself is perceived as such an in-group. In particular, it may build the trust among all parties in the corporate governance system by creating sound rules of the game (corporate governance politics and norms), strong control system for the rules implementation, and sound results of such control. The private property rights ignorance is to be eliminated by increasing transparency and enlarging the distinction between owners, managers, and other stakeholders, where third parties do not see company's assets as their own properties for self-enrichment. Finally, the strong compliance and anti-corruption policies must to be in place to avoid the well-spread practice of personal favors.

On the basis of the vast literature research and several sets of the empirical verification, authors of the following book built the set of indicators for sustainable corporate governance that is able to summarize the effort of different authors (Table 4.1).

The metrics were verified for comparability and value relevance on the international sample, where authors used similar metrics for measuring the sustainable governance indicators applying it to 110 listed companies from 7 countries. The calculations were as well adopted for the sample of 212 Russian largest listed companies with 424 company-year observations for the years 2019–2020. The hypothesis tested was about a significant and positive linear relationship between sustainable governance and market value (Cardoni et al., 2022).

Table 4.1 Core metrics of the sustainable governance

Core metrics and disclosures	Calculation
ESG strategy and risk management component	
Setting purpose	Company has a stated purpose linked to E, S and G
Integrating ESG risks and opportunities into business process	Description of the risks from E, S and G spheres Inclusion of emerging principal risks of fourth industrial revolution (data security/digitalization etc.)
Board composition and stakeholder engagement component	
Board composition	No COB-CEO duality Independence of the majority of the Board Gender diversity and/or membership of under-represented social groups Stakeholder representation in the board or committees
Impact of material issues on stakeholders	The material topics are identified Stakeholders were questioned during this process
Anti-corruption and business ethics component	
Anti-corruption	Governance body members have received training on the organization's anti-corruption policies and procedures Employees have received training on the organization's anti-corruption policies and procedures Business partners have received training on the organization's anti-corruption policies and procedures
Protected ethics advice and reporting mechanisms	Company has internal and external mechanisms for seeking advice about ethical behavior (politics, norms, standards, events, etc.) Company has internal and external mechanisms for reporting concerns about unethical behavior (hotline and alike)
Transparency	
Disclosure about integration of ESG risks and opportunities into business process	Disclosure of company-level ESG risk factors as opposed to generic sector risks Disclosure of the board appetite in respect of these risks Disclosure of how these risks have moved over time Disclosure of the response to risks
Board composition disclosure	Disclosure of tenure on the governance body Disclosure of the number of each individual's other significant positions and commitments and the nature of the commitments Disclosure of competencies relating to economic, environmental, and social topics (at least one)
Anti-corruption disclosure	Disclosure of total number and nature of incidents of corruption confirmed during the current year, but related to previous years Disclosure of total number and nature of incidents

(continued)

Table 4.1 (continued)

Core metrics and disclosures	Calculation
	of corruption confirmed during the current year, related to this year Disclosure of the number of employees/governance members or business partners

Sources: prepared by the authors

The names of the indicators were formulated with the help of the WEF 2020 White Paper "Measuring Stakeholder Capitalism" prepared by Deloitte, EY, KPMG, and PwC (World Economic Forum, 2020b). The indicators from the WEF Paper are oriented toward long-term value generation, consistent with a wide range of previous studies, and can be easily adapted by the companies.

4.2 Standard Setters at the Forefront of Sustainable Governance Measurement

The World Economic Forum (WEF) presented above is the most promising initiative in the field of ESG metrics development. WEF proposed to unify the approach to the definition of ESG metrics, including corporate governance in the context of ESG, in the document. The initiative of the World Economic Forum is the most successful in unifying the understanding of ESG, including G, since it has been developed:

- Together with more than 140 global business leaders.
- With consultations from more than 200 companies, investors, and other key players.
- In collaboration with Deloitte, EY, KPMG, and PwC.
- Using multiple ESG reporting frameworks (GRI, Sustainability Accounting Standards Board, Task Force on Climate-related Financial Disclosures).
- Keeping value creation as an aim of the governance.
- Taking into account current state of corporate governance development.

> Governance is foundational to achieving long-term value, by aligning and driving both financial and societal performance, as well as by ensuring accountability and building legitimacy with stakeholders (World Economic Forum, 2020a, b).

The World Economic Forum document identified five topics describing corporate governance in the context of ESG or good corporate governance. The topic of the priority goal (Governing Purpose) means that the company must have a stated goal related to environmental, social, and economic results. Risk and opportunity oversight is the explicit inclusion of ESG risks in corporate governance and related processes and projects. The Quality of Governing Body metrics show whether the activities of the board of directors/supervisory board and top management are aimed at long-term value creation and stakeholder participation, as well as whether the performance of management bodies in ESG aspects is monitored and whether it improves over time. The topic of stakeholder engagement is aimed at reflecting the regular process of identifying and selecting key stakeholders, analyzing their interests, and actively involving stakeholders in decision-making processes. Ethical behavior indicators show the difference between a formalist approach to the implementation of ethical principles and a strategic approach aimed at the long-term implementation of ESG goals (World Economic Forum, 2020a, b).

The International Sustainability Standards Board (ISSB) is another one organization that is aimed at boosting the comparability of ESG and governance data. ISSB has published the Exposure Draft IFRS S1 General Requirements for Disclosure of Sustainability-related Financial Information. These standards have potential to become the most widespread one because of the united forces of Value Reporting Foundation, IFRS, and some other bodies in the creation of this requirements. However, they are still in a draft form and under comments in 2022 and not ready for a widespread adoption (IFRS, 2022).

Similar to WEF framework, ISSB focuses on financial materiality of ESG data. In relation to governance data, the objective of such information is to enable investors and other users of financial statements to understand the corporation's governance processes, controls, and procedures in relation to sustainability-related risks and opportunities. As for the content of the information, company is expected to provide data on the governance body or bodies and their role in the oversight of ESG risks and opportunities and information about management's role in those processes. In particular, company shall present the data about the responsibility for ESG risks oversight, its skills and competencies, fixation of its responsibilities in terms of reference, board mandates and other related policies, the frequency of the meetings or other ways of communication about ESG initiatives, and risks and responsibilities status. Governance data shall also include the board decision on corporate strategy, major transactions, and its risk management policies, including any assessment of trade-offs and analysis of sensitivity to uncertainty. ESG is expected to be included into the corporate strategy with particular quantitative and qualitative targets in relation to significant sustainability-related risks and opportunities, linked to the remuneration policies and monitored regularly. Dedicated controls and procedures shall be applied to all the levels of company's hierarchical levels and integrated with other internal departments or functions (IFRS, 2022).

4.3 Sources of Information on Sustainable Governance

Information about sustainable corporate governance is considered to be a nonfinancial information, while the definition of nonfinancial disclosure is still controversial in literature, including a wide range of alternatives in terms of contents and frameworks.

About contents, the meaning of nonfinancial disclosure is still ambiguous and multifaceted (Tarquinio & Posadas, 2020). This term has often referred to information about society and the environment, though most academics define and understand nonfinancial disclosure differently, as corporate social responsibility (CSR) issues, intellectual capital, or information related to strategy. These definitions pave the way for conceptualizing nonfinancial disclosure as a genus and its different understandings (i.e., CSR, ESG information, strategy, etc.) as species. Nonfinancial disclosure is considered as an umbrella term that often overlaps other concepts or is deemed to be synonymous of different terms, not limiting to CSR and sustainability but also including value creation and strategy. With the advent of EU Directive 2014/95/UE, there has been a growing interest in nonfinancial disclosure. The number of publications on nonfinancial disclosure peaked in 2015, and it is gaining momentum across the European countries and large companies. This evidence makes the understanding of the term even more significant in Europe, where nonfinancial reporting has become mandatory for listed large entities exceeding 500 employees and, alternatively, 40 million euros turnover or 20 million euros total assets.

Sustainability and annual reports are the main sources of nonfinancial information. They are the most popular among companies for delivering the sustainability aims, ESG risks, and environmental or social performance to stakeholders (Du et al., 2017; Dumay & Hossain, 2019). Unlike other forms of corporate environmental and social communication, sustainability reports are of higher assurance: reports are always verified by the internal auditor, very often by the third-party auditor and often by the wide range of stakeholders through the mechanism of public hearings. Thus, these reports are indeed used by stakeholders and are more potential to affect a firm's market valuations in the short and long term (Du et al., 2017).

With respect to the frameworks, literature and practice have developed various positions over time.

A first position emphasizes the role of qualitative documents included in the annual report, such as management commentary. This supplementary information have gained momentum especially to disclose firm's strategy and business model. For example, in the UK, the Companies Act (2006) requires large listed companies to prepare the Strategic Report; the International Financial Reporting Standard (IFRS) has proposed guidelines for the presentation of the management commentary, now contained in the Practice Statement 1. With this approach, nonfinancial disclosure reports the description of the business, the objectives and strategies, the resources, the risks, and the relationships that characterize the company. A second position concerns the sustainability reports and the nonfinancial standards and frameworks which are based on reporting forms. They aim to help firms and other entities to understand, measure, and communicate the impact their activities can generate on the various dimensions of sustainability in the economic, environmental, and social dimensions.

Out of the reporting frameworks, there are several most widespread ones used by the majority of the companies. Moreover, the abovementioned WEF framework and ISSB initiative are largely based on the already developed infrastructure of nonfinancial reporting standards. Out of them are GRI standards, SASB, and TCFD.

Global Reporting Initiative (GRI) was established in 1997 as an international reporting standard for use by organizations reporting on sustainable development. The key principle of GRI is materiality, that is, determining which operations are considered essential and mandatory for disclosure in the report. Materiality is understood for a wide range of stakeholders and has been understood as a double materiality, including financial materiality and impact materiality. That includes information on economic value creation at the level of the reporting company for the benefit of investors (shareholders) and information on the reporting company's impact on the economy, environment, and people for the benefit of multiple stakeholders, such as investors, employees, customers, suppliers, and local communities (GRI, 2022). The corporate opinion on the materiality is considered to be in priority. As of 2022, the structure is modular with interconnected standards inside. Three series of standards support the reporting process: the GRI Universal Standards, which apply to all organizations; the GRI Sector Standards, applicable to specific sectors; and the GRI Topic Standards, each listing disclosures relevant to a particular topic.

Out of the advantages of using GRI standards is their international popularity and wide variety of the topics of the indicators that cover all spheres of triple-bottom line performance. Standard may be used by the large enterprises as well as smaller ones. However, there are some drawbacks. Firstly, it is not integrated into legislative systems of the countries; thus, it may contradict some laws, especially in the case of methodology for some particular indicators. This, in turn, leads to incorrect use of the standard by the companies and incorrect comparisons of these companies by investors. Moreover, standard does not provide the minimum of the indicators for compliance by companies and what reduces comparability and lowers the interest of SMEs in this standard.

Sustainability Accounting Standards Board (SASB) is another one framework that is now a part of International Financial Standard (IFRS) Foundation. Starting from 2011, the key principles of SASB standards are global applicability, financial materiality, and standard setting through industry-specific sustainability reporting.

It is vital to mention that the standards for the disclosure of sustainability-related financial information are being drafted by the IFRS Foundation—with which the newly established ISSB is charged with—and will be based on financial materiality for an investor audience only. As of 2022, the structure of the standards contains about 77 specialized guidelines for the disclosure of nonfinancial indicators.

Out of the main strong points of the standards is their substantial support for a large number of different industries—consumer goods, technology, services, transport, and others. Similar to GRI, the standard can be applied by organizations of any size, industry, and location. However, the standard is focused on investors, so the composition of indicators is revised based on the situation in the markets. The emphasis on financial significance leads to dependence on financial reporting standards.

Task Force on Climate-Related Financial Disclosures (TCFD) is a nonprofit organization that has been developing recommendations on voluntary disclosure of information about financial risks of companies arising from global climate change since 2017. The key principles of TCFD are the development of reporting that helps companies to provide stakeholders with information that can be used to attract investment and make the right management decisions. Among the strengths of the standards is a deep study of risks and their impact on the performance of companies. However, the standard is characterized by a concentration on problems related only to climate change.

The standards described above are the core frameworks for the companies even though they differ largely in their core conception of the ESG metrics.

The divergence of the standards is still a relevant issue for ESG data users. Fast change of the power distribution between the standard setters is changing the landscape significantly. For example, the ISSB creation almost gulped the Integrated Reporting Initiative audience, while GRI started to share the popularity with upcoming ISSB recommendations. Changes are expected to continue with the centralization unification trends in mind. Largely, such trends are empowered by investors and their request to limit the material list of ESG issues to the financially relevant ones, while other issues are expected to gravitate toward the best practice of the value-driven ESG aspects.

4.4 ESG Ratings on Sustainable Governance

The ratings in the field of ESG vary at a great extent. In 2019, there was a large debate on the ESG rating divergence that has been started by Berg et al. (2019) paper "Aggregate Confusion: The Divergence of ESG Ratings." Authors investigated the heterogeneity of ESG rating evaluations of corporate ESG performance based on data from six prominent ESG rating agencies as of 2019: KLD, Sustainalytics, Moody's ESG (Vigeo-Eiris), S&P Global (RobecoSAM), Refinitiv (Asset4), and MSCI.

The Pearson correlations of the ESG ratings are on average 0.54 and range from 0.38 to 0.71, while financial ratings are usually correlated at more than 90%. Out of the main reasons of the correlation is stated the difference in methodologies, when rating agencies measure the same attribute using different indicators. What is more, the rater affect is proven to be key in creating all kinds of divergences since raters tend to evaluate all indicators of one company in a similar way: if the first indicator takes on a high value, then rater tends to assign high values for all indicators instead of objective estimations of each of them (Berg et al., 2019).

Example 4.2 MSCI ESG rating methodology

One of the examples of the methodologies can be presented by the case of MSCI ESG rating—one of the most used rating by the investors in the field.

(continued)

Example 4.2 (continued)

Agency uses a bunch of sources to estimate the level of a company's ESG performance, including both corporate data and third-party data:

- Macro data of the segment or geographic level from academic, government, NGO datasets (e.g., Transparency International, US EPA, World Bank).
- Corporate disclosure (10-K, sustainability report, proxy report, AGM results, etc.)
- Governmental databases.
- 3400+ media sources
- NGO data.
- Other stakeholder sources regarding specific companies.

Some other rating agencies may also use questionnaires, while the practice of using public information is more common. Since ESG information is valued by stakeholders, the public availability of data is a key element of corporate responsibility. What is more, the third-party information, like mass media, may tell more about company ESG performance than company itself since it has not only shining version of the company profile but also represents the scandals, spills, bribery cases, minor conflicts, etc.

While evaluating the level of the ESG performance, MSCI estimates the following metrics: ownership characteristics, board and committee composition, pay figures, accounting metrics, policies and practices, business and geographic segments, and controversies. Deduction-based scoring model is applied for calculating E, S, and G values, while the total ESG score is derived by making the weighted average of underlying pillar scores of E, S, and G. After estimating company's score, the adjusted relative to industry peers is calculated, and final measure is later transformed into letter (e.g., A, B, C).

Although the data sources are mainly publicly available, communication is a vital stage in rating undertakings. MSCI reaches out to companies as part of data review processes.

Time periods of rating evaluations tend to shorten from yearly observations to a real-time tracking. MSCI is also corresponding to the following trend: companies are monitored on a systematic and ongoing basis, including daily monitoring of controversies and governance events. New information is reflected in reports on a weekly basis, and significant changes to scores trigger analyst review and re-rating (MSCI, 2022).

The divergence of ESG rating scores may be explained by the existence of the gaps in the ESG data market, where companies represent the supply side and investors represent the demand side.

The market of such investments is growing and forming the demand for company's valuation of ESG information. Assets managed by ESG funds equal to approximately USD2 trillion, and they tripled in 2010–2020, increased by 50% in

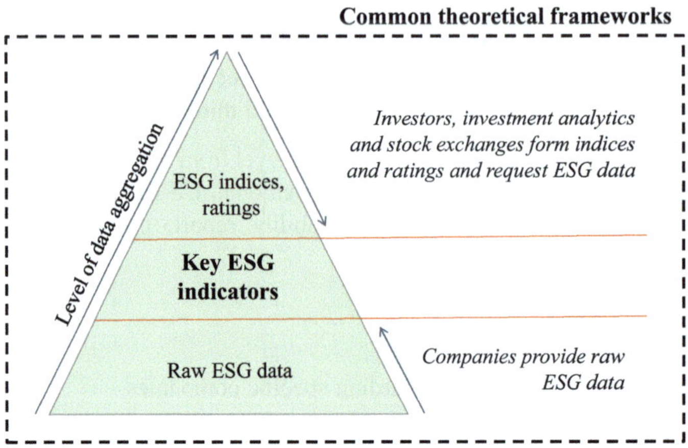

Fig. 4.1 Flow of ESG information. Source: constructed by the authors

2019, and doubled in 2020. What is more, responsible investments start to play a great role in an international arena: responsible investment assets account to 36% out of all professionally managed assets globally, including ESG funds (Adams & Abhayawansa, 2021).

Companies are trying to provide relevant ESG information to investors in order to be evaluated by responsible investors and taken into their portfolio. Two-thirds out of 5200 world leading companies from 52 countries use GRI standards in their reports; more than 2500 companies publish integrated reports using IR approach; 1400+ use SASB standards (GRI Secretariat, 2021; International Integrated Reporting Council (IIRC), 2021; Sustainability Accounting Standards Board, 2021).

Thus, ESG information flow goes from two sides: investors are forming the request for aggregated and comparable ESG data across organizations, while companies are providing a wide list of raw ESG data. This flow of ESG information may be presented in the form of triangle with the most aggregated type of data (in forms of ESG ratings and indices) on the top and the least aggregated one (raw ESG data) in the bottom (Fig. 4.1).

However, the gap between provided raw information and requested aggregated one exists. Companies provide limited ESG information or heterogeneous data with nonstandardized methodologies, metrics, and measures or even nonmaterial ESG data: what creates a lack of relevant ESG information for investors (Cardoni et al., 2019). At the same time, investors are not sure which indicators they need for their valuation models.

The disability for the demand and supply to meet arises mainly because of the fact that there is *no understanding from both sides on which ESG information they do actually need* to analyze or perform. It means that there is no clear vision of the common framework that would clarify the list of key ESG indicators that are material both for investors (for value estimation) and for companies (for measuring and publishing necessary data).

Investors lack the theoretical frameworks for understanding how to integrate ESG issues into existing valuation models and how the company value is changing under the influence of ESG challenges. The classical understanding of the value generated by the company from using the access to limited or unique resources (Friedman, 2007) is becoming less popular in the modern agenda of environmental awareness, social instability, economic damages from COVID-19 outbreak, and geopolitical risks of 2022. Greater role is devoted to the value that is generated by company together with a wide range of stakeholders, including shareholders and investors, and that is distributed between these stakeholders (Freeman, 2010; Hart & Milstein, 2003; Wheeler et al., 2003). However, the most used framework for describing such value—sustainable value—has not been revised for more than 15 years after a widely cited paper by Hart and Milstein was published in 2003 (Hart & Milstein, 2003; Cardoni et al., 2020).

Companies need to be aware of how to generate the value for investors by effective integration of ESG issues. Existing theoretical frameworks provide variety of approaches:

- Considering a meta-management approach to sustainability (Asif et al., 2011).
- Concentrating at company's identity and organizational culture (Eccles et al., 2012).
- Applying different strategies for integrating each ESG issue at each maturity level of sustainability practice (Baumgartner & Ebner, 2010).
- Using a purchasing, operations, and supply chain management contexts for ESG implementation (Whitelock, 2019).
- Other.

However, all these frameworks are criticized for being too abstract, lacking the detailed explanations for particular E, S, and G issues, and having a scarce of practical examples on the basis of real enterprises (Aguinis & Glavas, 2012; Despeisse et al., 2013; McWilliams & Siegel, 2011; Rosen & Kishawy, 2012).

Finally, the interconnection between the ESG and the value is vital to investigate for both companies and investors: nowadays, empirical research on this topic is as popular as it is controversial. A lot of meta-studies over a worldwide sample of companies conclude that a positive relationship between ESG and financial performance might be expected, and some authors reveal mixed result or even evidences of negative dependence (Cardoni et al., 2022). Research over developing countries brings even more divergent results than in the worldwide research (Crotty, 2016; Zhao, 2012).

Sustainable governance tends to be the framework that may overcome the problem of disperse ESG issues. The use of the most strategically oriented and widely disclosed metrics for the research helps the concept to overcome the main reasons of inconsistency in scientific research: starting from the lack of cross-company comparability of ESG data and finishing with conceptual focus on too broad list of stakeholders and too many aspects of ESG (Cardoni et al., 2019; Orlitzky, 2015). Governance is the most operationalized block out of E, S, and G, thus providing companies with bunch of indicators for evaluation and reporting.

Finally, governance is a meta-construct that is a precondition and a result of ESG, which means that the development of sustainable governance leads to the consistent multiplicity positive effects on environmental and social business activities as well as economic performance and value of the company.

References

Adams, C. A., & Abhayawansa, S. (2021). Connecting the COVID-19 pandemic, environmental, social and governance (ESG) investing and calls for 'harmonisation' of sustainability reporting. *Critical Perspectives on Accounting, 102309*. https://doi.org/10.1016/j.cpa.2021.102309

Aguinis, H., & Glavas, A. (2012). What we know and don't know about corporate social responsibility: A review and research agenda. *Journal of Management, 38*(4), 932–968. https://doi.org/10.1177/0149206311436079

Allegrini, M., & D'Onza, G. (2011). Corporate governance, risk management e responsabilità sociale fra presente e futuro dell'attività di internal auditing. *Management Control, 1*, 151–178.

Asif, M., Searcy, C., Zutshi, A., & Ahmad, N. (2011). An integrated management systems approach to corporate sustainability. *European Business Review, 23*(4), 353–367. https://doi.org/10.1108/09555341111145744

Baumgartner, R. J., & Ebner, D. (2010). Corporate sustainability strategies: Sustainability profiles and maturity levels. *Sustainable Development, 18*(2), 76–89. https://doi.org/10.1002/sd.447

Berg, F., Kölbel, J. F., & Rigobon, R. (2019). *Aggregate confusion: The divergence of ESG ratings* (SSRN scholarly paper no. 3438533). https://doi.org/10.2139/ssrn.3438533

Cardoni, A., Kiseleva, E., Arduini, S., & Terzani, S. (2022). From sustainable value to shareholder value: The impact of sustainable governance and anti-corruption programs on market valuation. *Business Strategy and the Environment*. https://doi.org/10.1002/bse.3328

Cardoni, A., Kiseleva, E., & Terzani, S. (2019). Evaluating the intra-industry comparability of sustainability reports: The case of the oil and gas industry. *Sustainability, 11*(4), 1093. https://doi.org/10.3390/su11041093

Cardoni, A., Kiseleva, E., & Taticchi, P. (2020). In search of sustainable value: A structured literature review. *Sustainability, 12*(2), 2. https://doi.org/10.3390/su12020615

Crotty, J. (2016). Corporate social responsibility in The Russian Federation: A contextualized approach. *Business & Society, 55*(6), 825–853. https://doi.org/10.1177/0007650314561965

Despeisse, M., Oates, M. R., & Ball, P. D. (2013). Sustainable manufacturing tactics and cross-functional factory modelling. *Journal of Cleaner Production, 42*, 31–41. https://doi.org/10.1016/j.jclepro.2012.11.008

Dicuonzo, G., Donofrio, F., Iannuzzi, A. P., & Dell'Atti, V. (2022). The integration of sustainability in corporate governance systems: An innovative framework applied to the European systematically important banks. *International Journal of Disclosure and Governance, 19*(3), 249–263. https://doi.org/10.1057/s41310-021-00140-2

Du, S., Yu, K., Bhattacharya, C. B., & Sen, S. (2017). The business case for sustainability reporting: Evidence from stock market reactions. *Journal of Public Policy & Marketing, 36*(2), 313–330. https://doi.org/10.1509/jppm.16.112

Dumay, J., & Hossain, M. A. (2019). Sustainability risk disclosure practices of listed companies in Australia. *Australian Accounting Review, 29*(2), 343–359. https://doi.org/10.1111/auar.12240

Eccles, R. G., Perkins, K. M., & Serafeim, G. (2012). How to become a sustainable company. *MIT Sloan Management Review, 53*(4), 43-+.

European Commission. (2022). Proposal for a Directive of the European Parliament and of the Council on Corporate Sustainability Due Diligence and amending Directive (EU) 2019/1937. Retrieved May 6, 2023, from https://eur-lex.europa.eu/resource.html?uri=cellar:bc4dcea4-9584-11ec-b4e4-01aa75ed71a1.0001.02/DOC_1&format=PDF

Freeman, R. E. (2010, March). Strategic management by R. Edward Freeman. Cambridge Core. https://doi.org/10.1017/CBO9781139192675

Friedman, M. (2007). The social responsibility of business is to increase its profits. In W. C. Zimmerli, M. Holzinger, & K. Richter (Eds.), *Corporate ethics and corporate governance* (pp. 173–178). Springer. https://doi.org/10.1007/978-3-540-70818-6_14

GRI. (2022). *The materiality madness: Why definitions matter* (Issue 3). Retrieved May 6, 2023, from https://www.globalreporting.org/media/r2oojx53/gri-perspective-the-materiality-madness.pdf

GRI Secretariat. (2021). GRI - Standards. https://www.globalreporting.org/standards/

Hart, S. L., & Milstein, M. B. (2003). Creating sustainable value. *Academy of Management Perspectives, 17*(2), 56–67. https://doi.org/10.5465/ame.2003.10025194

Hussain, N., Rigoni, U., & Orij, R. P. (2018). Corporate governance and sustainability performance: Analysis of triple bottom line performance. *Journal of Business Ethics, 149*(2), 411–432. https://doi.org/10.1007/s10551-016-3099-5

IFRS. (2022). *Exposure Draft and comment letters: General Sustainability-related Disclosures*. Retrieved May 6, 2023, from https://www.ifrs.org/projects/work-plan/general-sustainability-related-disclosures/exposure-draft-and-comment-letters/

International Integrated Reporting Council (IIRC). (2021). *International <IR> Framework*. Retrieved May 6, 2023, from https://integratedreporting.org/resource/international-ir-framework/

Jacoby, G., Liu, M., Wang, Y., Wu, Z., & Zhang, Y. (2019). Corporate governance, external control, and environmental information transparency: Evidence from emerging markets. *Journal of International Financial Markets, Institutions and Money, 58*, 269–283. https://doi.org/10.1016/j.intfin.2018.11.015

Lu, L. W. (2021). The moderating effect of corporate governance on the relationship between corporate sustainability performance and corporate financial performance. *International Journal of Disclosure and Governance, 18*(3), 193–206. https://doi.org/10.1057/s41310-020-00099-6

Mccarthy, D. J., & Puffer, S. M. (2008). Interpreting the ethicality of corporate governance decisions in Russia: Utilizing integrative social contracts theory to evaluate the relevance of agency theory norms. *Academy of Management Review, 33*(1), 11–31. https://doi.org/10.5465/amr.2008.27745006

McWilliams, A., & Siegel, D. S. (2011). Creating and capturing value: Strategic corporate social responsibility, resource-based theory, and sustainable competitive advantage. *Journal of Management, 37*(5), 1480–1495. https://doi.org/10.1177/0149206310385696

MSCI. (2022). *ESG ratings methodology*. Retrieved May 6, 2023, from https://www.msci.com/esg-and-climate-methodologies

Naciti, V., Cesaroni, F., & Pulejo, L. (2022). Corporate governance and sustainability: A review of the existing literature. *Journal of Management and Governance, 26*(1), 55–74. https://doi.org/10.1007/s10997-020-09554-6

Orlitzky, M. (2015). The politics of corporate social responsibility or: Why Milton Friedman has been right all along. *Annals in Social Responsibility, 1*(1), 5–29. https://doi.org/10.1108/ASR-06-2015-0004

Peng, X., & Zhang, R. (2022). Corporate governance, environmental sustainability performance, and normative isomorphic force of national culture. *Environmental Science and Pollution Research, 29*(22), 33443–33473. https://doi.org/10.1007/s11356-022-18603-6

Rosen, M. A., & Kishawy, H. A. (2012). Sustainable manufacturing and design: Concepts, practices and needs. *Sustainability, 4*(2), 154–174. https://doi.org/10.3390/su4020154

Sánchez, J. L. F., Sotorrío, L. L., & Díez, E. B. (2011). The relationship between corporate governance and corporate social behavior: A structural equation model analysis. *Corporate Social Responsibility and Environmental Management, 18*(2), 91–101. https://doi.org/10.1002/csr.244

Sustainability Accounting Standards Board. (2021, December 17). *Value reporting foundation*. SASB standards. SASB. Retrieved May 6, 2023, from https://www.sasb.org/

Tarquinio, L., & Posadas, S. C. (2020). Exploring the term "non-financial information": An academics' view. *Meditari Accountancy Research, 28*(5), 727–749. https://doi.org/10.1108/MEDAR-11-2019-0602

Wheeler, D., Colbert, B., & Freeman, R. E. (2003). Focusing on value: Reconciling corporate social responsibility, sustainability and a stakeholder approach in a network world. *Journal of General Management, 28*(3), 1–28. https://doi.org/10.1177/030630700302800301

Whitelock, V. G. (2019). Multidimensional environmental social governance sustainability framework: Integration, using a purchasing, operations, and supply chain management context. *Sustainable Development, 27*, 923. https://doi.org/10.1002/sd.1951

World Economic Forum. (2020a). *Toward Common Metrics and Consistent Reporting of Sustainable Value Creation*. Retrieved May 6, 2023, from https://www.weforum.org/whitepapers/toward-common-metrics-and-consistent-reporting-of-sustainable-value-creation/

World Economic Forum. (2020b). *Measuring Stakeholder Capitalism: Towards Common Metrics and Consistent Reporting of Sustainable Value Creation*. Retrieved May 6, 2023, from https://www.weforum.org/reports/measuring-stakeholder-capitalism-towards-common-metrics-and-consistent-reporting-of-sustainable-value-creation/

Zhao, M. (2012). CSR-based political legitimacy strategy: Managing the state by doing good in China and Russia. *Journal of Business Ethics, 111*(4), 439–460. https://doi.org/10.1007/s10551-012-1209-6

Chapter 5
The Role of Sustainable Governance in Value Generation of the Company

5.1 Theoretical Propositions

Going forward from the discussion on the gaps in the ESG data market, we claim that sustainable governance concept is able to provide both companies and investors with the common framework as well as set of comparable metrics or indicators for evaluating company's performance. However, the relevant question arises of whether these sustainable governance metrics lead to sustainable financial performance that is still the prior aim of corporate development.

To begin with, in the empirical research of the ESG, financial performance connection is extremely controversial. A lot of meta-studies over worldwide sample of companies conclude that a positive relationship between ESG and financial performance might be expected (Griffin & Mahon, 1997; Margolis et al., 2007; Orlitzky et al., 2003), and some authors reveal mixed results (Aouadi & Marsat, 2018; Friede et al., 2015) or even evidences of negative dependence (Riley et al., 2003; Rodrigo et al., 2016).

Governance role in financial performance has much more homogeneity in results than total ESG scores, but the sustainable or stakeholder characteristics of the governance are usually not considered or not verified separately out of the bulk of financial data.

Several reasons may be the cause of the ambiguities in the results.

First and foremost, it is the lack of cross-company comparability of ESG data (Amaeshi & Grayson, 2009; Cardoni et al., 2019). The globalized trend for ESG codification and standardization is overcoming the first problem by increasing comparability between companies, industries, and even countries. However, standardized ESG management may increase costs for the company without a contribution to a greater economic performance (Cardoni et al., 2020; Ioannou & Serafeim, 2019; Orlitzky, 2015; Porter & Kramer, 2006).

The other counterproductive aspect of ESG research is the conceptual focus on a broad list of stakeholders with the need to satisfy all requests of all relevant

A. Cardoni, E. Kiseleva, *Sustainable Governance*, CSR, Sustainability, Ethics & Governance, https://doi.org/10.1007/978-3-031-37492-0_5

stakeholders or to meet the collective needs (Orlitzky, 2015; Orlitzky et al., 2003), while there is a call in the literature for narrower concepts and variables (Orlitzky, 2015).

We claim that sustainable corporate governance may be a metric that both allows to measure the value of ESG undertakings and overcome the abovementioned disadvantages of ESG metrics. In the context of ESG issues, the G part may be called ethical governance or sustainable corporate governance and determined as a corporate governance system based on the integration of shareholders' and other stakeholders' expectations, which safeguards society and the environment.

Institutional and stakeholder theories support our claim. Theories postulate that good governance influences financial results.

According to basic *agency theory* (Hölmstrom, 1979; Jensen & Meckling, 1976; Ross, 1973), the relationship among shareholders and the board of directors is considered an agency relationship in which shareholders delegate the realization of corporate activities to the board of directors, whereas the actions or reactions of the boards of directors underline the information asymmetry between firm managers and financial markets (Eisenhardt, 1989).

The agency theory of the firm suggests that when individuals engage in firm relationships, they are utility maximizers, self-seeking, and opportunistic, and, therefore, the governance system introduces mechanisms that align the interests of principals (owners) with those of their agents (the mangers) (Aguilera, 2005; Jensen & Meckling, 1976). Thus, the main argument for examining the relative importance of governance in market value is based on the argument that monitoring mechanisms are able to reduce agency problems between owners and managers and to reduce costs related to this problem (Caixe & Krauter, 2014; Fu, 2019; Worokinasih, 2020). In this sense, the vast literature proposes the evidence of a relationship between better governance practices and higher market value (Ammann et al., 2011; Black et al., 2006; Caixe & Krauter, 2014; Durnev & Kim, 2005; Worokinasih, 2020).

Institutional theory (DiMaggio & Powell, 1983; Meyer & Rowan, 1977; Tolbert & Zucker, 1999) is supposed to be more relevant for the modern economic systems (Aguilera, 2005). Institutional theory is defined as a set of formal and informal rules that affect business activity (North, 2010).

According to institutional theory, the corporate organization is a social system or collective entity with pluralist interests and some common goals (Aguilera, 2005). A firm's right to exist is legitimized if its value system is consistent with that of the larger social system of which it is part of but threatened when there is actual or potential conflict between the two value systems (Suchman, 1995). According to this paradigm, corporate governance helps in incorporating external norms and rules into operations and structures that enables the legitimacy and social acceptance (Briano-Turrent & Rodríguez-Ariza, 2016; DiMaggio & Powell, 1983). This ability, in turn, makes corporate governance a vital factor in economic growth and financial stability (Aguilera, 2005).

The stakeholder theory (Donaldson & Preston, 1995; Freeman, 1984; Mitchell et al., 1997) with its interpretation of corporate governance is gaining the largest attention nowadays (Aguilera, 2005). According to stakeholder theory, companies should design their corporate strategies considering the interests of the key stakeholders, while those stakeholders are considered to be contributors to company's wealth creation or destruction. Some authors even state that the main purpose of the company becomes to build the value for its stakeholders.

According to stakeholder theory, corporate governance is responsible not only for protecting and increasing shareholder wealth but also for ensuring that strategic decisions are beneficial for all other stakeholders (Allais et al., 2017; Shahzad et al., 2016). Such stakeholder-oriented or sustainable corporate governance focuses on the value-creating relationships with all stakeholders, avoiding negative impacts on the environment and society (Lombardi et al., 2019; Sánchez et al., 2011). Moreover, the extension of corporate governance with the firm's key stakeholders is expected to be guided by more than just regulatory or legal requirements to encompass an ethical dimension (Lombardi et al., 2019). Stakeholder orientation creates a competitive advantage for proactive companies both from the possibility of being rewarded by the market and from avoiding risks (Allais et al., 2017).

5.2 Business-Level Interpretation

On the business level, the vital role of sustainable governance is also supported by the practical peculiarities of the concept, described below.

To begin with, sustainable governance includes such elements as ESG strategy, ESG risk management, board composition that ensures the realization of the rights of shareholders and other stakeholders, developed anti-corruption policies, and business ethics practices (World Economic Forum, 2020). By this, sustainable governance becomes a meta-construct that is a *precondition and a result of ESG*, which means that the development of sustainable governance leads to the consistent multiplicated positive effects on environmental and social business activities as well as economic performance (Jacoby et al., 2019).

Sustainable governance factor better represents the needed *strategic long-term approach to ESG factors* (Cardoni et al., 2020). Resource-based theory postulates that organizational activities, resources, or capabilities cannot lead to competitive advantage if they are not unique to a firm (Barney, 1991; Orlitzky, 2015), while strategic approach to ESG is the only one that is able to generate a unique ESG initiative road map for a corporation and not to assimilate to an average industry-related practice. Thus, sustainable governance is the main driver for ESG initiatives to become a competitive advantage rather than additional irrelevant cost. High E or S performance does not have such impacts without proper sustainable governance: they may increase costs for the company without a contribution to a greater economic performance (Cardoni et al., 2020; Ioannou & Serafeim, 2019; Orlitzky, 2015; Porter & Kramer, 2006).

Overview 5.1 The role of strategic approach to ESG
Hristov et al. in their recent research of 2022 supported the role of strategic approach to ESG. Authors analyzed the answers of the 85 managers specialized in sustainable corporate strategy. Interviewees offered an in-depth discussion on how the integration between sustainability and strategy brings advantages. Managers supported four main performance outcomes of strategic ESG: financial performance, image and reputation, stakeholder perception, and cultural change (Hristov et al., 2022).

The *data on corporate governance* are the least controversial and most common for different industries and countries due to its simplicity in calculation (i.e., number of independent directors or percentage of women onboard) in comparison with all ESG data. Additionally, governance data are widely published in annual report unlike E or S data that are usually located solely in separate and less widespread sustainability reports. Such peculiarity helps to overcome the lack of cross-company comparability of ESG data (Amaeshi & Grayson, 2009) that is accepted to be the largest challenge in integrating ESG information into investors' evaluation processes (Amel-Zadeh & Serafeim, 2018; Cardoni et al., 2019).

Sustainable governance enables *stronger connection to financial measures* comparing to other ESG issues. ESG concept is defined too widely with its three completely different and very broad issues of economic, social, and governance development (Schramade, 2016) with not clear criteria for choosing and weighting of ESG indicators (Busch et al., 2016). Sustainable governance is a narrower concept with shareholder, or investor, as the main interested party that eliminated the difficulties in multidimensional nature of ESG. At the same time, it is still oriented at responding wide stakeholder requests due to its meta-characteristics and multiplicity effect on all ESG issues.

At last, governance indicators *may be incorporated into the existing company's value models*: they are closer by nature to financial indicators and are more homogeneous unlike social or environmental subunits. Thus, sustainable governance may play a key role in comprehending the interactions between financial markets and corporate ESG strategies (Crifo et al., 2019) and in helping to identify valuable business from less valuable one (Lo, 2010).

References

Aguilera, R. V. (2005). Corporate governance and director accountability: An institutional comparative perspective. *British Journal of Management, 16*(s1), 39–53. https://doi.org/10.1111/j.1467-8551.2005.00446.x

Allais, R., Roucoules, L., & Reyes, T. (2017). Governance maturity grid: A transition method for integrating sustainability into companies. *Journal of Cleaner Production, 140*, 213–226. https://doi.org/10.1016/j.jclepro.2016.02.069

Amaeshi, K., & Grayson, D. (2009). *The challenges of mainstreaming environmental, social and governance (ESG) issues in investment decisions* (A contribution to the valuing business project, p. 53) [A mini-survey of practitioners' reports]. http://citeseerx.ist.psu.edu/viewdoc/download?doi=10.1.1.466.1773&rep=rep1&type=pdf

Amel-Zadeh, A., & Serafeim, G. (2018). Why and how investors use ESG information: Evidence from a global survey. *Financial Analysts Journal, 74*(3), 87–103. https://doi.org/10.2469/faj.v74.n3.2

Ammann, M., Oesch, D., & Schmid, M. M. (2011). Corporate governance and firm value: International evidence. *Journal of Empirical Finance, 18*(1), 36–55. https://doi.org/10.1016/j.jempfin.2010.10.003

Aouadi, A., & Marsat, S. (2018). Do ESG controversies matter for firm value? Evidence from international data. *Journal of Business Ethics, 151*(4), 1027–1047. https://doi.org/10.1007/s10551-016-3213-8

Barney, J. (1991). Firm resources and sustained competitive advantage. *Journal of Management, 17*(1), 99–120. https://doi.org/10.1177/014920639101700108

Black, B. S., Jang, H., & Kim, W. (2006). Does corporate governance predict firms' market values? Evidence from Korea. *The Journal of Law, Economics, and Organization, 22*(2), 366–413. https://doi.org/10.1093/jleo/ewj018

Briano-Turrent, G. d. C., & Rodríguez-Ariza, L. (2016). Corporate governance ratings on listed companies: An institutional perspective in Latin America. *European Journal of Management and Business Economics, 25*(2), 63–75. https://doi.org/10.1016/j.redeen.2016.01.001

Busch, T., Bauer, R., & Orlitzky, M. (2016). Sustainable development and financial markets: Old paths and new avenues. *Business & Society, 55*(3), 303–329. https://doi.org/10.1177/0007650315570701

Caixe, D. F., & Krauter, E. (2014). The relation between corporate governance and market value: Mitigating endogeneity problems. *Brazilian Business Review, 11*(1), 80–110. https://doi.org/10.15728/bbr.2014.11.1.5

Cardoni, A., Kiseleva, E., & Lombardi, R. (2020). A sustainable governance model to prevent corporate corruption: Integrating anticorruption practices, corporate strategy and business processes. *Business Strategy and the Environment, 29*(3), 1173–1185. https://doi.org/10.1002/bse.2424

Cardoni, A., Kiseleva, E., & Terzani, S. (2019). Evaluating the intra-industry comparability of sustainability reports: The case of the oil and gas industry. *Sustainability, 11*(4), 1093. https://doi.org/10.3390/su11041093

Crifo, P., Escrig-Olmedo, E., & Mottis, N. (2019). Corporate governance as a key driver of corporate sustainability in France: The role of board members and investor relations. *Journal of Business Ethics, 159*(4), 1127–1146. https://doi.org/10.1007/s10551-018-3866-6

DiMaggio, P. J., & Powell, W. W. (1983). The iron cage revisited: Institutional isomorphism and collective rationality in organizational fields. *American Sociological Review*, *48*(2), 147–160. JSTOR. https://doi.org/10.2307/2095101

Donaldson, T., & Preston, L. E. (1995). The stakeholder theory of the corporation: Concepts, evidence, and implications. *The Academy of Management Review*, *20*(1), 65–91. JSTOR. https://doi.org/10.2307/258887

Durnev, A., & Kim, E. H. (2005). To steal or not to steal: Firm attributes, legal environment, and valuation. *The Journal of Finance, 60*(3), 1461–1493. https://doi.org/10.1111/j.1540-6261.2005.00767.x

Eisenhardt, K. M. (1989). Agency theory: An assessment and review. *Academy of Management Review, 14*(1), 57–74. https://doi.org/10.5465/amr.1989.4279003

Freeman, R. (1984). *Strategic management: A stakeholder approach*. Cambridge University Press.

Friede, G., Busch, T., & Bassen, A. (2015). ESG and financial performance: Aggregated evidence from more than 2000 empirical studies. *Journal of Sustainable Finance & Investment, 5*(4), 210–233. https://doi.org/10.1080/20430795.2015.1118917

Fu, Y. (2019). The value of corporate governance: Evidence from the Chinese anti-corruption campaign. *The North American Journal of Economics and Finance, 47*, 461–476. https://doi.org/10.1016/j.najef.2018.06.001

Griffin, J. J., & Mahon, J. F. (1997). The corporate social performance and corporate financial performance debate: Twenty-five years of incomparable research. *Business & Society, 36*(1), 5–31. https://doi.org/10.1177/000765039703600102

Hölmstrom, B. (1979). Moral Hazard and observability. *The Bell Journal of Economics, 10*(1), 74–91. JSTOR. https://doi.org/10.2307/3003320

Hristov, I., Chirico, A., & Ranalli, F. (2022). Corporate strategies oriented towards sustainable governance: Advantages, managerial practices and main challenges. *Journal of Management and Governance, 26*(1), 75–97. https://doi.org/10.1007/s10997-021-09581-x

Ioannou, I., & Serafeim, G. (2019). Corporate sustainability: A strategy? *Harvard Business School*, SSRN Journal. https://doi.org/10.2139/ssrn.3312191

Jacoby, G., Liu, M., Wang, Y., Wu, Z., & Zhang, Y. (2019). Corporate governance, external control, and environmental information transparency: Evidence from emerging markets. *Journal of International Financial Markets, Institutions and Money, 58*, 269–283. https://doi.org/10.1016/j.intfin.2018.11.015

Jensen, M. C., & Meckling, W. H. (1976). Theory of the firm: Managerial behavior, agency costs and ownership structure. *Journal of Financial Economics, 3*(4), 305–360. https://doi.org/10.1016/0304-405X(76)90026-X

Lo, S.-F. (2010). Performance evaluation for sustainable business: A profitability and marketability framework. *Corporate Social Responsibility and Environmental Management, 17*(6), 311–319. https://doi.org/10.1002/csr.214

Lombardi, R., Trequattrini, R., Cuozzo, B., & Cano-Rubio, M. (2019). Corporate corruption prevention, sustainable governance and legislation: First exploratory evidence from the Italian scenario. *Journal of Cleaner Production, 217*, 666–675. https://doi.org/10.1016/j.jclepro.2019.01.214

Margolis, J. D., Elfenbein, H. A., & Walsh, J. P. (2007). Does it pay to be good? A meta-analysis and redirection of research on the relationship between corporate social and financial performance 1001 (48109–1234). https://citeseerx.ist.psu.edu/viewdoc/download?doi=10.1.1.424.7803&rep=rep1&type=pdf

Meyer, J. W., & Rowan, B. (1977). Institutionalized organizations: Formal structure as myth and ceremony. *American Journal of Sociology, 83*(2), 340–363. JSTOR.

Mitchell, R. K., Agle, B. R., & Wood, D. J. (1997). Toward a theory of stakeholder identification and salience: Defining the principle of who and what really counts. *Academy of Management Review, 22*(4), 853–886. https://doi.org/10.5465/amr.1997.9711022105

North, D. C. (2010). *Understanding the process of economic change*. https://press.princeton.edu/books/paperback/9780691145952/understanding-the-process-of-economic-change

Orlitzky, M. (2015). The politics of corporate social responsibility or: Why Milton Friedman has been right all along. *Annals in Social Responsibility, 1*(1), 5–29. https://doi.org/10.1108/ASR-06-2015-0004

Orlitzky, M., Schmidt, F. L., & Rynes, S. L. (2003). Corporate social and financial performance: A meta-analysis. *Organization Studies, 24*(3), 403–441. https://doi.org/10.1177/0170840603024003910

Porter, M. E., & Kramer, M. R. (2006). Strategy and society: The link between competitive advantage and corporate social responsibility. *Harvard Business Review, 23*(5), sd.2007.05623ead.006. https://doi.org/10.1108/sd.2007.05623ead.006

Riley, R. A., Pearson, T. A., & Trompeter, G. (2003). The value relevance of non-financial performance variables and accounting information: The case of the airline industry. *Journal of Accounting and Public Policy, 22*(3), 231–254. https://doi.org/10.1016/S0278-4254(03)00021-8

Rodrigo, P., Duran, I. J., & Arenas, D. (2016). Does it really pay to be good, everywhere? A first step to understand the corporate social and financial performance link in Latin American controversial industries. *Business Ethics: A European Review, 25*(3), 286–309. https://doi.org/10.1111/beer.12119

Ross, S. A. (1973). The economic theory of agency: The Principal's problem. *The American Economic Review*, *63*(2), 134–139. JSTOR.

Sánchez, J. L. F., Sotorrío, L. L., & Díez, E. B. (2011). The relationship between corporate governance and corporate social behavior: A structural equation model analysis. *Corporate Social Responsibility and Environmental Management, 18*(2), 91–101. https://doi.org/10.1002/csr.244

Schramade, W. (2016). Integrating ESG into valuation models and investment decisions: The value-driver adjustment approach. *Journal of Sustainable Finance & Investment, 6*(2), 95–111. https://doi.org/10.1080/20430795.2016.1176425

Shahzad, A. M., Rutherford, M. A., & Sharfman, M. P. (2016). Stakeholder-centric governance and corporate social performance: A cross-National Study. *Corporate Social Responsibility and Environmental Management, 23*(2), 100–112. https://doi.org/10.1002/csr.1368

Suchman, M. C. (1995). Managing legitimacy: Strategic and institutional approaches. *Academy of Management Review, 20*(3), 571–610. https://doi.org/10.5465/amr.1995.9508080331

Tolbert, P. S., & Zucker, L. G. (1999). The institutionalization of institutional theory. In *Studying organization: Theory & method* (pp. 169–184). Sage. https://doi.org/10.4135/9781446218556

World Economic Forum. (2020). *Toward common metrics and consistent reporting of sustainable value creation*. Retrieved October 20, 2022, from https://www.weforum.org/whitepapers/toward-common-metrics-and-consistent-reporting-of-sustainable-value-creation/

Worokinasih, S. (2020). The Mediating Role of Corporate Social Responsibility (CSR) Disclosure on Good Corporate Governance (GCG) and Firm Value. 14, 10. Retrieved May 06, 2023, from https://core.ac.uk/download/pdf/287817575.pdf

Chapter 6
Sustainable Governance in Specific Settings of Listed and Unlisted Companies, SMEs, and Family Businesses

6.1 From Corporate Governance to Sustainable Governance: The Broader Contextualization of the New Approach

The evolution of the governance concept toward a sustainability perspective generated a significant extension of the different contexts of implementation.

The traditional approach to corporate governance historically characterized a context of large listed companies, operating in developed countries with financial systems mainly based on financial markets, typical of the Anglo-Saxon tradition (Bottenberg et al., 2017). As already underlined in the previous chapters, the theoretical framework was mainly based on the agency theory which interpreted the strategic orientation of the companies according to a shareholder approach, consistent with the financial market and large listed companies' mechanisms.

Additionally, considering the teaching principle in management and business schools in the late nineteenth century, scholars and institutional investors praised the shareholder primacy system highly (Zhang et al., 2022a). The theoretical discussion and the regulation of the corporate governance concerned almost exclusively these contexts and the research, both qualitative and quantitative, focused mainly on the characteristics, determinants, and implications of the corporate governance for large listed companies (Zhang et al., 2022a).

Still in such context, at the beginning of the new millennium, the corporate governance experienced a dramatic evolution toward a dimension of greater ethics and responsible behavior, through a profound review of the operating mechanisms of the market and of the companies. Corporate scandals and the great pressure from stakeholders force managers to abandon the conventional approaches that only focus on maximizing profits (Deeg, 2009; Adnan & Tandigalla, 2017), creating a growing tendency to adopt stakeholder-oriented models to attain sustainable development (Yoshikawa et al., 2021).

A. Cardoni, E. Kiseleva, *Sustainable Governance*, CSR, Sustainability, Ethics & Governance, https://doi.org/10.1007/978-3-031-37492-0_6

In all the developed economies, the authorities have started reforms regarding the functioning of the markets and the principles of corporate governance for listed companies. Simultaneously, the stronger attention and sensitivity to social and environmental issues coming from operators and consumers increasingly developed an orientation toward the issues of corporate social responsibility (CSR). In such evolution, the financial markets and the increasing quotation of smaller companies in specific segments assigned to corporate governance a fundamental role in assuring an institutional and social legitimization. By reevaluating the ethical dimension of the businesses and the need to orient the strategy in the long-term value creation, literature has moved to emphasize the relevance of the corporate governance according to the pillars of the legitimacy theory (Masocha & Fatoki, 2018). Governance has thus become a tool for maintaining or enhancing the corporate legitimacy in society.

A new approach to corporate governance has been embedded into the regulatory systems, influencing the listed companies' behavior through various forms of isomorphic pressures (Di Maggio & Powell, 1983). The concept began to overcome the exclusive perimeter of the large listed companies to involve the organizational and strategic settings of listed SMEs. Any company operating in the regulated markets needed to set up governance mechanisms to be consistent, on the one hand, to the regulatory framework and, on the other, to the new issues of sustainable finance coming from financial markets. Public, government, shareholders, and investors appreciated these requirements as a professional and efficient behavior, although the actual benefits of having such a committee might be insignificant.

With the advent of sustainability era, the panorama has further changed, and the contextualization of governance has become even broader, both from a theoretical and regulatory point of view.

The growing influence of corporate social responsibility and the need to integrate economic performance with environmental and social outcomes have profoundly influenced the entrepreneurial orientation not only in the financial markets but in the broader scenario of the business environment. Stakeholder engagement and integration of sustainable value into the corporate strategies have become fundamental pillars valid for each type of company, regardless of the status of listed companies and also including SMEs and small family businesses.

The importance of ESG factors, although originally developed in the context of regulated markets, is increasingly affecting smaller companies, accompanied by the growing pressure deriving from the institutional and banking regulations. Looking at the European context, it is possible to mention the reform of credit lending and monitoring (EBA, 2020) and the recently issued Directive on Sustainability Reports (EU, 2022). Also in the economies of emerging countries, there is a visible attention to investments and sustainable finance specifically focused to SMEs (Fatoki, 2021), confirming a global tendency to consider the sustainability perspective valid for each kind of company. The need to integrate the traditional governance structures with an orientation toward sustainability has become universal, which made the interpretation of the strategy and governance orientation according to the stakeholder theory even more relevant.

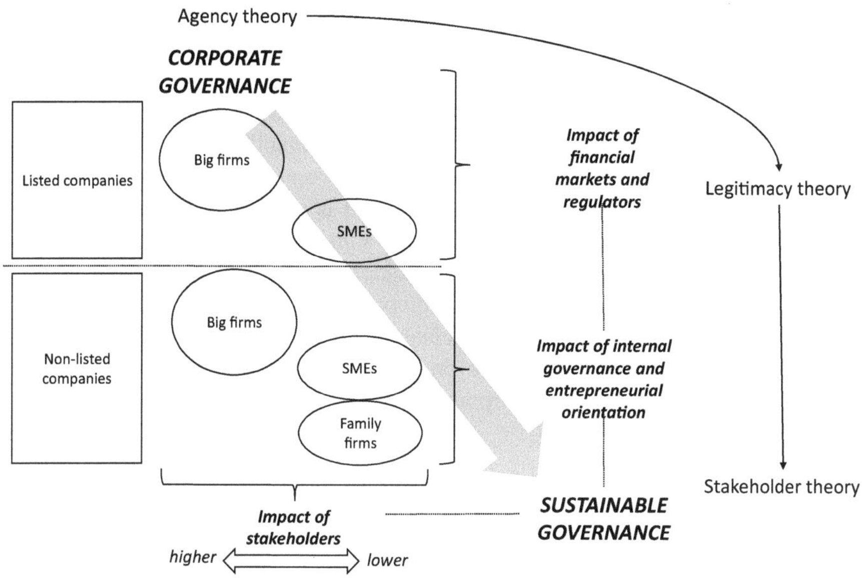

Fig. 6.1 The broader inclusion of contexts in the shift from corporate governance to sustainable governance. Source: Authors' elaboration

The evolutionary path just described can be represented as follows (Fig. 6.1):

Summarizing, the concept of sustainable governance presents some common pillars and components that need to be interpreted and tailored taking into account the specific context of application. Under the same inspirational principles, the elements of differentiation may concern the status of listed or not, the smaller size, the characteristics of industry, or the geographical context (developed or emerging countries). All these dimensions can influence the concrete definition of sustainable governance, in many cases interacting through a multidimensional impact.

The analysis developed above allows to emphasize two main implications.

First, taking into account the concept development, the sustainable governance can be considered a multi-theoretical approach. Especially in the context of the larger companies that are fully challenged by the three issues (agency, legitimacy, and stakeholder), the challenging goal in the practical design and implementation is to find the right balance of different perspectives that in some cases can mutually reinforce. For example, literature arguments a sort of convergence between legitimacy and stakeholder theory. Particularly, Romero et al. (2018) develop a multi-theoretical approach in which there is a clear link between the two theories. In large companies (Romero et al., 2018; Calace, 2014), the motivations underlying the attention to stakeholder and the need to be legitimized in the society tend to mutually reinforce with positive effect on the quality of disclosure (Cormier & Magnan, 2015; Crossley et al., 2021).

Second, moving to consider the unlisted companies and SMEs, the impact of agency problem is less perceived, and legitimacy forces are mostly operating through the coercive and normative pressures (Masocha & Fatoki, 2018). In such context, legitimacy and stakeholder theory can easily develop opposing tensions (Romero et al., 2018; Cormier & Magnan, 2015) with a significant impact on the sustainable governance and transparency. According to stakeholder theory, firms apply a "substantive management approach" to sustainability, able to generate real changes in process or practice as improvement in governance processes to respond to stakeholders' demand. With legitimacy theory, it is a more "symbolic managerial approach," mainly aimed to enhance company's image. As effectively expressed by Amran et al. (2015), while a genuine stakeholder management is driven by external demand, legitimacy theory focuses on a company's own internal needs to secure its legitimacy to operate in society. In such latent trade-off, if the governance approach is driven only by a type of coercive pressure, the impact of the change remains limited, without permeating the corporate strategy and culture.

6.2 The Context of Unlisted Companies and SMEs: The Growing Relevance of Sustainable Governance at Global Level

The traditional association of corporate governance with larger listed companies is mainly due to the following reasons. Firstly, the agency problem in large companies mostly refers to the relationships between shareholders (owners) and managers. However, in many SMEs, the owner is the sole proprietor and manager (Hart, 1995), so the separation of ownership and management is not as present in SMEs as it is with larger firms. This makes some people argue that corporate governance should not apply to SMEs. Secondly, the question of accountability by SMEs to the public is nonexistent since they do not depend on public funds. Most especially the sole proprietorship businesses do not necessarily need to comply with any disclosure requirement and hence are in little or no need of corporate governance principles (Abor & Adjasi, 2007a, b). Thirdly, SMEs have few employees—who are mostly relatives of the owner, and for that matter, no separation between ownership and control, and there is no need for corporate governance in their operations.

Despite all these arguments, since the last two decades, the issue of governance has been increasingly studied also with reference to unlisted companies and SMEs, both in emerging countries (Zhong, 2015) and developed economies (Popoola, 2022). Literature has broadly demonstrated that the adoption of corporate governance principles by unlisted SMEs represents a necessary step in their entrepreneurial orientation and strategy alignment with the business environment. As affirmed by Bill Witherell, Head OECD Directorate for Financial, Fiscal, and Enterprise Affairs, good corporate governance underpins market confidence, integrity, and efficiency and hence promotes economic growth and financial stability (Whiterell, 2002). Indeed, a growing body of literature emphasizes that good governance can be really beneficial for any kind of company (Zhang & Wei, 2021). External board members bring into the firm expertise and knowledge on financing options available and strategies to source such finances, thus dealing with the credit constraint problem of SMEs. They are able to challenge strategies by management (Dzomonda, 2022) and are thus able to inject better management practices to attract resources. Small unlisted firms that excel in corporate governance tend to have higher chances of attracting funding for their businesses than those with weak corporate governance practices (Zumente & Bistrova, 2021). Essentially, excelling in corporate governance principles helps businesses to establish long-term relationships with key stakeholders such as banks, customers, investors, and shareholders (Zhang et al., 2022a). This can help resource-constrained SMEs to unlock resources among their key stakeholders (Dzomonda, 2022) and impact positively on long-term survival (Popoola, 2022).

Corporate governance has implications for economic development especially in helping to increase the flow of financial capital to firms in developing countries. When fully implemented, good corporate governance would ensure that SMEs are well-run in order to earn the confidence of investors and lenders (Radebe, 2017). Among several factors influencing good corporate governance for SMEs, the banking and financial perspective has the higher impact to demonstrate the relevance of the issue for the economic growth at large (King & Levine, 1993). A number of authors have also linked SME governance to financial performance (Nasrallah & El Khoury, 2022), also supporting its relevance especially in countries with weak legal protections for investors (Klapper & Love, 2004).

The universal value of good corporate governance is also confirmed in the managerial perspective (McLarty & O'Dowd, 2023). During the last decades, several guidance documents have been written focusing on the specific governance needs of SMEs. It is possible to mention several examples (IFC, 2019) represented by the following:

1. *Corporate Governance Guidance and Principles for Unlisted Companies in Europe* prepared by the European Confederation of Director Associations (ecoDa) in 2010. The document uses a phased approach, making a distinction between the use of a basic framework that applies to all companies, including the smaller and less complex organizations and then more sophisticated measures for larger and complex organizations.
2. *Governance in SMEs—A Guide to the Application of Corporate Governance in Small and Medium Enterprises*, a Southern-African Institute of Directors (IoDSA) guide that contains a section specifying how listed, larger company guidance can be interpreted and applied by SMEs.
3. *Guidelines on Corporate Governance for SMEs in Hong Kong Written by the Institute of Director (HKIoD, 2014).* The document aims to account for the heterogeneity of SMEs by varying recommendations depending on key company characteristics, most commonly size, organizational complexity, and shareholding structure.

In this scenario, the strategic relevance assumed by sustainability makes the SME sustainable governance of paramount importance for several reasons. First is due to the sizeable share of SMEs in the total number of companies around the world (over 90%), playing a fundamental role in the global economy for the value production and employment. At the same time, as highly demonstrated (Crossley et al., 2021), SMEs cause significant effects to the environment and wider society, thereby raising major ethical, social, and environmental dilemmas for stakeholders, such us governments, regulators, and activists. Second is due to SMEs facing specific challenges when attempting to integrate sustainability reporting (Rossi & Luque-Vílchez, 2020; Massa et al., 2015). Third is due to increased stakeholder pressure to orient SMEs toward sustainable business practices (Massa et al., 2015) and business model sustainability. Finally, SMEs are mostly exposed to the negative effect of crises (Zutshi et al., 2021) and discontinuities related to responsible investments (Sun et al., 2022), and the attention of stakeholders is even stronger. A particular attention has been devoted to sustainable finance for SMEs, considering their strategic relevance to the achievement of the sustainable development goals (SDGs). Even sustainable finance has tended to focus on large firms, while current research demonstrates (Fatoki, 2021) that sustainable finance is an emerging trend in small business finance.

Literature has recently focused on the integration of sustainability principles into SMEs (Jayeola, 2015), highlighting a relationship that remains problematic (Cardoni et al., 2022). In large companies, both listed and unlisted, the sustainability orientation has been operational for years, and the different players are gradually aligning their professional standard to the new paradigm (Lai & Stacchezzini, 2021). Even if certain critical issues tend to persist, such as green washing or an exclusively formal approach, awareness has now grown at all levels: institutions, companies, consultants, auditors, investors, and financial markets.

The significant changes in the global economic environment are requiring major improvements in the governance for SMEs led by an entrepreneur, often characterized by poor management and inadequate governance practices, provoking a high rate of failure. As demonstrated by the framework presented in the following box, the evolution of governance must accompany each stage of development of the company (IFC, 2019). Looking at the future perspective, for the smaller firms, the evolution must start from the fundamental pillars of classical corporate governance and create the preconditions for an advancement to sustainable governance.

Overview 6.1 A governance framework for SME sustainability and growth proposed by the International Finance Corporation (IFC) Corporate Governance Group

The International Finance Corporation (IFC), a member of the World Bank Group, advances economic development and improves the lives of people by encouraging the growth of the private sector in developing countries. The commitment of IFC is to support the World Bank Group's twin goals of ending extreme poverty and boosting shared prosperity. The IFC Corporate

(continued)

Overview 6.1 (continued)
Governance Group helps support corporate governance improvements and reform efforts in emerging markets and developing countries while leveraging and integrating knowledge tools, expertise, and networks at the global and regional levels.

In the paper "Governance for SMEs Sustainability and Growth" (IFC, 2019), the position expressed is very clear: as small and medium enterprises (SMEs) evolve from start-up to maturity, they need to "grow into governance." The paper proposes a governance model designed to support the organic growth of small and mid-size businesses by identifying issues and recommendation that are fit for purpose through each stage of their evolution (Table 6.1). The proposed framework would aim to improve small business company culture and guide family-owned businesses through the process of formalizing family engagement through bodies such as family councils.

The framework is developed along two dimensions:

1. Components of the governance, represented by "culture and commitment to good governance" (*policies, processes, and organizational structure*), "decision-making and strategic oversight" (*decision-making policies and bodies, leadership style*), "risk governance and internal control" (*internal checks and balance*), "disclosure and transparency" (*communication with internal and external stakeholder*), and "ownership" (*founders/shareholders/family*).

2. SMEs' stage of growth, including the following stages: "start-up" where product/service development and market testing are the first priorities"; "active growth," whose defining features are the need for growth through sales, people, and increasing complexity; "organizational development" through specialization, professional policies, structures, and staff; "business expansion," with additional capital that is often required and a possible increase in the number of shareholders that necessitates more formality in the corporate governance arrangements. At this stage, the company's governance starts taking on the characteristics of "classic corporate governance," including a board of directors.

In the sustainable governance of SMEs, particular focus should be devoted to family businesses. As can be seen in the framework above, the role of the founder and of the family members may be particularly relevant to strengthen the governance mechanisms in the various stages of development.

The issue of corporate governance in family businesses has been extensively addressed in the literature (Siebels & Knyphausen-Aufseß, 2012). Family businesses tend to be complex because, in addition to dealing with common business opportunities and requirements, they must consider the needs and desires of the owner family (Esteban et al., 2006).

In terms of governance, family businesses may present several benefits (Ward, 2002). Family leaders remain in their positions for longer periods, a fact that implies

Table 6.1 Governance framework for SME sustainability and growth

Governance components	Stage of growth			
	Start-up	Active growth	Organizational development	Business expansion
Culture and commitment to good governance	Small multitasking team; high degree of informality; few systems, established "on the go"	Distinct functions and organizational structure start emerging; simple systems to enable functions to collaborate	Increased professionalization of functions; formalizing organizational structure, policies, and procedures	Continuation of trends started in the previous stage
Decision-making and strategic oversight	Highly centralized decision-making by the founder(s); autocratic leadership style	Emergence of delegation to management; consultative leadership style—Largely autocratic but with input from key managers and advisers	Professional managers are hired; Decentralization of authority through division/functional management Collaborative management style	Separation of strategic and operational decision-making; institutional decision-making style, based on defined organizational structure, roles, and procedures
Risk governance and internal controls	Founders are fully involved in opera- tions—Limited need for checks and balances	Introducing internal controls to support delegation of authority	Detailing authorities and accountability; Systems are formalized and automated; Developing practices to control main operational risks	Focus on proactive and strategic risk management
Disclosure and transparency	Everyone knows everything	Silos—Good within, but challenging between silos Basic external information shared on products offered	Internally: Improving cross-divisional/functional information sharing; enhanced external business-related information	Optimizing communication between management, board, and shareholders
Ownership	Single owner or couple of individuals; Founders personally control every aspect of business	New minority share- holders possible (internal or related); Founders remain dominant and fully engaged; increasing number of family members becoming	New minority share- holders possible (internal or related); New investors informally influence strategy but are not directly involved in operations; if a major investor enters,	Common options: a. Founders, private equity, and other investors b. Growing family ownership/ generational change c. Go public (IPO)

(continued)

Table 6.1 (continued)

Governance components	Stage of growth			
	Start-up	Active growth	Organizational development	Business expansion
		involved in operations	company moves to next stage	Investors require tools for control and direction of the company

Source: adapted from IFC, 2019

added business continuity and stability (Arcari, 2018). Also, beyond profitability, family businesses aim at continuity and prudence and exercise disciplined growth. Finally, they have more loyal and long-lasting employees and executives, as a result of their long-term relationship with the family. On the other hand, the typical governance in family business can create overlap, confusion, and possible disagreement between family and business needs. Additionally, the sole leadership of the entrepreneur and its relatives can limit the attraction of external resources (Siebels & Knyphausen-Aufseß, 2012) and the potentiality for the firms' strategic development. To this extant, a critical point in the firm's life cycle is the correct planning of the transition process (Sgrò et al., 2022; Esteban et al., 2006) and the decision to set a board of directors (Brenes et al., 2011), able to mediate among the needs of family shareholders, employees, other stakeholders, and the business (Sgrò et al., 2022).

In this scenario, the sustainable governance implementation remains problematic (Mariani et al., 2021; Cardoni et al., 2022). Even in such perspective, recent literature highlights the strategic relevance of setting up a board of directors with professional members (IFC, 2019) and the crucial role played by succession plan (Delmas & Gergaud, 2014) as effective strategies to remove the obstacles to make possible the long-term integration with the sustainability.

6.3 The Case of Institutional and Regulatory Changes in the European Context: The Impact for SMEs' Sustainable Governance

Two important institutional and regulatory changes are expected to greatly impact in the near future on the sustainable governance of European unlisted companies and SMEs. First is the reform of credit lending and monitoring promoted by the European Banking Authority (EBA, 2020). Second is the very recently issued Directive on Sustainability Reports (EU Directive 2022/2464).

Banks are the most important source of external funds especially for loans (Mishkin, 2001), and ensuring the flow of credit to SME sector provides a sound ground for economic growth (Cuevas et al., 1993). The persistent challenge to assure the adequate availability of finance is a universal and important problem among unlisted companies and SMEs at global level (Sprenger & Lazareva, 2022). SMEs' inability to access loans is often due to prescreening techniques employed by banks that include feasibility studies and collateral and track record (Adalessossi & Utku, 2015). It is highlighted that in most developing countries, lending to small businesses and entrepreneurs remains limited because financial intermediaries are apprehensive about supplying credit to businesses due to their high risk, small portfolios, and high transaction costs (Cuevas et al., 1993; Adalessossi & Utku, 2015). The banking system has always played a fundamental impact for SME's finance and economic growth that is now even more relevant with the advent of "green banking" (Zhang et al., 2022b), a banking approach whereby banks take the initiative to act as a fundamental player to stimulate the implementation of responsible investments and the sustainability strategy in the society.

In this scenario, a relevant reform has been introduced in the European context. In May 2020, the European Banking Authority (EBA) issued the new guidelines on credit lending and monitoring (EBA, 2020). Among the various innovations, these measures contemplate the integration of ESG factors in credit policies by all European banks, with clear consequences for the credit access conditions for SMEs in the near future. The integration of ESG factors concerns both the strategic and organizational profile of the banking institutions as well as their credit lending and monitoring policies. More specifically, it is established that the banking institutions "should incorporate ESG factors and associated risks in their credit risk appetite and risk management policies, credit risk policies and procedures, adopting a holistic approach" (Art. 56- ABE/GL/2020/06). Moreover, in the next article, it is specified that "institutions should take into account the risks associated with ESG factors on the financial conditions of borrowers and in particular the potential impact of environmental factors and climate change, in their credit risk appetite, policies, and procedures" (Art. 57—ABE/GL/2020/06). The integration of ESG factors is also explicitly mentioned in the sections concerning credit lending to micro- and small enterprises (5.2.5) and medium-sized and large enterprises (5.2.6).

Emphasis is then placed on ESG factors, which must be integrated into the three pillars of prudential supervision with the aim of countering the high exposure of the credit system to nonperforming loans and improving credit quality. During the stress test, the ECB will assess the level of sustainability present in the credit institutions and the greater or lesser active participation of ESG factors in credit intermediation activity.

Consequently, banks will be required to request information relating to the environmental, social, and governance factors present in the business and planning profiles of all companies, assessing the degree of integration of these three areas with the economic-financial performance in order to improve the risk/return profile of loans. The capacity to manage the new banking requirements will greatly depend on sustainable governance orientation.

The second significant change has recently occurred. In April 2021, the European Commission developed a proposal (Corporate Sustainability Directive Proposal 2021/0104) to extend mandatory nonfinancial reporting to all listed SMEs, with the exception of listed microenterprises. Furthermore, the Commission clarified that "the transition to a sustainable economy is likely to mean that collecting and sharing sustainability information becomes common business practice for companies of all sizes. Therefore, in parallel to the new rules proposed today for large companies, the Commission is also proposing the development of separate, proportionate standards for SMEs."

The proposal has definitively approved with the Directive (EU Directive 2022/2464), providing for the extension of the sustainability disclosure obligation also to.

- Listed companies with fewer than 500 employees, including also SMEs (small and medium-sized enterprises) whose securities are traded on European regulated markets.
- Large companies, even if not listed.

Consequently, even large private or unlisted family businesses will have to communicate to stakeholders' projects, processes, impacts, results, risks in environmental and social matters, as well as the main aspects connected with the governance of sustainability.

On the other hand, the nonfinancial disclosure will not be an obligation for SMEs, unless they are listed. The unlisted SMEs may opt for the voluntary preparation of sustainability reports on the basis of consistent and proportionate standards, through independent documents or disclosure in the Management Report.

Summarizing, in the short run, the European Union will have its own sustainability reporting standards on all ESG issues, based on a multi-stakeholder perspective and of both a generic ("sector agnostic") and sector specific ("sector specific" nature). The European Directive reaffirms the principle of "double materiality," according to which information, in order to be material, must be relevant to the company from an economic, financial, and the socio-environmental perspective. In this sense, a distinction is made between "financial materiality" and "impact materiality" which are both equally important for European sustainability reporting.

The EFRAG (European Financial Reporting Advisory Group), an advisory body of the European Commission, has been mandated by the Directive to issue differentiated European sustainability reporting standards for SMEs. The proposed standards will be transferred to the European Commission for analysis and evaluation before their issue. So far 13 draft standards have been developed, 2 of which are transversal, 5 on environmental issues, 4 on social issues, and 2 on governance issues.

In the recent approval of the final version of the Directive and the first set of standards with reference to the governance subject, only one has been confirmed, namely, the one referring to "business conduct."

The reporting requirements will be phased in over time for different kinds of companies. The first companies will have to apply the standards in financial year 2024, for reports published in 2025. Listed SMEs are obliged to report as from 2026, with a further possibility of voluntary opt out until 2028, and will be able to report according to separate, proportionate standards that EFRAG will develop next year.

In the meantime, EFRAG is working to prepare a definitive set of standards for sustainability reports to adopt on a voluntary by unlisted SMEs (EFRAG, 2022-EU

Voluntary Sustainability Reporting Standard for non-listed SMEs that are outside the scope of CSRD) that will follow in several extents the same inspirational principles of the ESRS already issued. In this perspective, given the strategic relevance of the EFRAG approach for European unlisted companies and SMEs, the focus reported below illustrates the broader components of sustainable governance defined in the full set of standards actually issued.

Overview 6.2 The EFRAG components of sustainable governance to disclose in sustainability reports

In the draft version of European Sustainability Reporting Standards (ESRS) defined for consultation, the issues of governance and its disclosure are addressed in ESRS 2, with reference to the general aspects, and in ESRS G1 and ESRS G2, specifically devoted to the operational aspects.

In the introductory part, *ESRS 2* defines the disclosure requirements for the "description by the undertaking of its governance and organization in relation to sustainability matters." In this standard, there are five disclosure requirements:

1. Roles and responsibilities of its governance bodies and management levels with regard to sustainability matters. The entity has to "provide an understanding of the distribution of sustainability-related roles and responsibilities throughout the undertaking's organization, from its administrative, management, and supervisory bodies to its executive and operational levels, the expertise of its governance."

2. Administrative, management, and supervisory bodies about sustainability matters. The entity has to "provide an understanding of how governance bodies and management level senior executives are informed about sustainability-related facts, decisions, and/or concerns that are within their responsibility so that they can effectively perform their duties in that respect."

3. Sustainability matters addressed by the undertaking's administrative, management, and supervisory bodies. The entity has to "provide information on whether the administrative, management, and supervisory bodies were adequately informed of the material sustainability-related impacts, risks, and opportunities arising or developing during the reporting period."

4. Integration of sustainability strategies and performance in incentive schemes. The entity has to "provide an understanding of how members of the administrative, management, and supervisory bodies are incentivized to properly manage the undertaking's sustainability impacts, risks, and opportunities and, along with other employees, to take steps toward implementing the sustainability strategy of the undertaking."

5. Statement on due diligence. "In light of its holistic nature which is covered by a number of crosscutting ESRS and topical disclosure requirements, the undertaking shall provide a mapping that reconciles the main

(continued)

Overview 6.2 (continued)
aspects of sustainability due diligence to the relevant disclosures in its sustainability statements."

The *ESRS G1*, which has not received definitive approval, dealt with "governance, risk management, and internal control." In this standard, the components to be disclosed were as follows: governance structure and composition; corporate governance code or policy; nomination process; diversity policy; evaluation process; remuneration policy; risk management processes; internal control processes; composition of the administrative, management, and supervisory bodies; and meetings and attendance rate.

The *ESRS G2* deals with "business conduct." Also in this case, there are ten components to be disclosed:

1. Business conduct culture. The entity has to "provide an understanding of how the administrative, management, and supervisory bodies are involved in forming, monitoring, promoting, and assessing the business conduct culture."

2. Policies and targets on business conduct. The entity has to "provide an understanding of the undertaking's ability (i) to mitigate any negative impacts and maximize positive impacts related to business conduct throughout its value chain and (ii) to monitor and manage the related risks.

3. Prevention and detection of corruption and bribery. The entity has to "provide transparency on the key procedures of the undertaking to prevent and detect, investigate, and respond to corruption or bribery-related incidents or allegations."

4. Anticompetitive behavior prevention and detection. The entity has to "provide transparency on the key procedures of the undertaking to prevent and detect, investigate, and respond to allegations or incidents of anticompetitive behavior."

5. Anti-corruption and anti-bribery training. The entity has to "provide an understanding of the undertaking's training and educational initiatives to develop and maintain awareness related to anti-corruption or anti-bribery and business conduct within the undertaking as well as in the value chain."

6. Corruption or bribery events. The entity has to "provide transparency on legal proceedings relating to corruption or bribery incidents during the reporting period and the related outcomes."

7. Anticompetitive behavior events. The entity has to "provide transparency on publicly announced investigations into or litigation concerning possible anticompetitive behavior of the undertakings that are ongoing during the reporting period."

8. Beneficial ownership. The entity has to "provide transparency on the individuals who ultimately own or control the undertaking's organizational and control structure, including beneficial owners."

(continued)

Overview 6.2 (continued)

9. Political engagement and lobbying activities. The entity has to "provide transparency on the types, purpose, and cost of political contributions and lobbying activities of the undertaking during the reporting period."

10. Payment practices. The entity has to "provide insights on the contractual payment terms and the average actual payments."

The standards for SME would not be legally binding but aim to be consistent with the CSRD by applying the ESRS disclosure principles with the highest degree of proportionality and simplification. Consistent with the conceptual approach and architecture of the draft ESRS, the actual version of standards for SME describes the reporting requirements related to the strategy, the business model, and the topical matters for governance, environment, and social responsibility of the undertaking, representing a first building block in the full disclosure requirements reported above for listed firms.

References

Abor, J., & Adjasi, C. K. D. (2007a). Corporate governance and the small and medium enterprises sector: Theory and implications. Corporate governance. *The International Journal of Business in Society, 7*, 111–122. https://doi.org/10.1108/14720700710739769

Abor, J., & Adjasi, C. K. D. (2007b). Corporate governance and the small and medium enterprises sector: Theory and implications. *The International Journal of Business in Society, 7*(2), 111–122.

Adalessossi, K., & Utku, B. D. (2015). Financing difficulties of small and medium enterprises (SMEs) in the West African economic and monetary union (WAEMU) area. *China-USA Business Review, 14*(2), 79–99.

Adnan, A. T. M., & Tandigalla, H. (2017). The dramatic shift in emphasis from a shareholder-dominated approach to a stakeholder-oriented corporate governance model. *European Journal of Business and Economics, 2017*, 12.

Amran, A., Ooi, S. K., Mydin, R. T., & Devi, S. S. (2015). The impact of business strategies on online sustainability disclosures. *Business Strategy and the Environment, 24*, 551–564. https://doi.org/10.1002/bse.1837

Arcari, A. M. (2018). Preventing crises and managing turnaround processes in SMEs. *The Role of Economic Measurement Tools, Management Control, 3*, 131–155. https://doi.org/10.3280/MACO2018-003007

Bottenberg, K., Tuschke, A., & Flickinger, M. (2017). Corporate governance between shareholder and stakeholder orientation: Lessons from Germany. *Journal of Management Inquiry, 2017*(26), 165–180.

Brenes, E. R., Madrigal, K., & Requena, B. (2011). Corporate governance and family business performance. *Journal of Business Research, 64*(2011), 280–285.

Calace, D. (2014). Non-financial reporting in Italian SMEs: An exploratory study on strategic and cultural motivations. *International Journal of Business Administration, 5*(3), 34–48.

Cardoni, A., Kiseleva, E., & Bellucci, A. (2022). The quality of SMEs stakeholder communication during strategic crises: The case of Italian unlisted SMEs. *Business Strategy and the Environment*, 1–17. https://doi.org/10.1002/bse.3300

Cormier, D., & Magnan, M. (2015). The economic relevance of environmental disclosure and its impact on corporate legitimacy: An empirical investigation. *Business Strategy and the Environment, 24*, 431–450. https://doi.org/10.1002/bse.1829

Crossley, R. M., Elmagrhi, M. H., & Ntim, C. G. (2021). Sustainability and legitimacy theory: The case of sustainable social and environmental practices of small and medium-sized enterprises. *Business Strategy and the Environment, 1*(23), 3740–3762. https://doi.org/10.1002/bse.2837

Cuevas, C., Fafchamps, M., Hanson, R., Moll, P., & Srivastava, P. (1993). *Case studies of enterprise finance in Ghana*. Final report, Regional Program on Enterprise Development, The World Bank.

Deeg, R. (2009). The rise of internal capitalist diversity? Changing patterns of finance and corporate governance in Europe. *Economies et Societes, 2009*(38), 552–579.

Delmas, M. A., & Gergaud, O. (2014). Sustainable certification for future generations: The case of family business. *Family Business Review, 27*(3), 228.

Di Maggio, P. J., & Powell, W. W. (1983). The iron cage revisited: Institutional isomorphism and collective rationality in organisational fields. *American Sociological Review, 1983*(48), 147–160.

Dzomonda, O. (2022). Environmental sustainability commitment and access to finance by small and medium enterprises: The role of financial performance and corporate governance. *Sustainability, 14*, 8863. https://doi.org/10.3390/su14148863

EBA. (2020). *EBA/GL/2020/06 - Guidelines on loan origination and monitoring*. Retrieved May 06, 2023, from https://www.eba.europa.eu/regulation-and-policy/credit-risk/guidelines-on-loan-origination-and-monitoring

Esteban, B., Kryssia, M., & German, M. (2006). Family business structure and succession: Critical topics in Latin American experience. *Journal of Business Research, 59*(3), 372–374.

EU (European Union). (2022). *Directive on Sustainability Reports (CSRD) 2022/2464*. Retrieved May 06, 2023, from https://eur-lex.europa.eu/legal-content/EN/TXT/?uri=CELEX:32022L2464

Fatoki, O. (2021). Sustainable finance and small, medium and micro enterprises in South Africa. *Academy of Accounting and Financial Studies Journal, 25*, 1–7.

Hart O. (1995). *Firms, contracts and financial structure*. Oxford University Press.

International Finance Corporation (IFC). (2019). *Governance for SMEs sustainability and growth*. Retrieved May 06, 2023, from https://www.ifc.org/wps/wcm/connect/topics_ext_content/ifc_external_corporate_site/ifc+cg/resources/private+sector+opinion/governance+for+sme+sustainability+and+growth

Jayeola, O. (2015). The impact of environmental sustainability practice on the financial performance of SMEs: A study of some selected SMEs in Sussex. *International Journal of Business Management and Economic Research, 2015*(6), 214–230.

King, R. G., & Levine, R. (1993). Finance, entrepreneurship and growth. *Journal of Monetary Economics, 32*, 1–30.

Klapper, L. F., & Love, I. (2004). Corporate governance, investor protection, and performance in emerging markets. *Journal of Corporate Finance, 10*(5), 703–728.

Lai, A., & Stacchezzini, R. (2021). Organisational and professional challenges amid the evolution of sustainability reporting: A theoretical framework and an agenda for future research. *Meditari Accountancy Research, 29*(3), 405–429. https://doi.org/10.1108/MEDAR-02-2021-1199

Mariani, M. M., Al-Sultan, K., & De Massis, A. (2021). Corporate social responsibility in family firms: A systematic literature review. *Journal of Small Business Management, 61*, 1192. https://doi.org/10.1080/00472778.2021.1955122

Masocha, R., & Fatoki, O. (2018). The impact of coercive pressures on sustainability practices of small businesses in South Africa. *Sustainability, 10*, 3032.

Massa, L., Farneti, F., & Scappini, B. (2015). Developing a sustainability report in a small to medium enterprise: Process and consequences. *Meditari Accountancy Research, 23*(1), 62–91. https://doi.org/10.1108/MEDAR-02-2014-0030

McLarty, R., O'Dowd, S. (2023). *Even small businesses need corporate governance. Management in practice*. Ideas from the Yale School of Management. Retrieved January 13, 2023, from https://insights.som.yale.edu/insights/even-small-businesses-need-corporate-governance

Mishkin, F. S. (2001). The economics of money. In *Banking and financial markets* (p. 186, 6th ed.). Addison Wesley.

Nasrallah, N., & El Khoury, R. (2022). Is corporate governance a good predictor of SMEs financial performance? Evidence from developing countries (the case of Lebanon). *Journal of Sustainable Finance and Investment, 2022*(12), 13–43.

Popoola, O. (2022). The impact of corporate governance on long term survival of small businesses in Canada. *Journal of the International Council for Small Business, 2022*, 1–9.

Radebe, M. S. (2017). The benefits of good corporate governance to small and medium enterprises (SMEs) in South Africa: A view on top 20 and bottom 20 JSE listed companies. *Problems and Perspectives in Management, 2017*(15), 271–279.

Romero, S., Ruiz, S., & Fernandez-Feijoo, B. (2018). Sustainability reporting and stakeholder engagement in Spain: Different instruments, different quality. *Business Strategy and the Environment, 28*, 221–232. https://doi.org/10.1002/bse.2251

Rossi, A., & Luque-Vílchez, M. (2020). The implementation of sustainability reporting in a small and medium enterprise and the emergence of integrated thinking. *Meditari Accountancy Research, 29*(4), 966–984. https://doi.org/10.1108/MEDAR-02-2020-0706

Sgrò, F., Palazzi, F., & Ciambotti, M. (2022). Business continuity and planning effectiveness: An empirical analysis of Italian manufacturing SMEs. *Management Control, 3*, 89–108. https://doi.org/10.3280/MACO2022-003005

Siebels, J. F., & Knyphausen-Aufseß, D. (2012). A review of theory in family business research: The implications for corporate governance. *International Journal of Management Reviews, 14*(3), 280–304.

Sprenger, C., & Lazareva, O. (2022). Corporate governance and investment-cash flow sensitivity: Evidence from Russian unlisted firms. *Journal of Comparative Economics, 2022*(50), 71–100.

Sun, J., Wang, F., Yin, H., & Zhao, R. (2022). Death or rebirth? How small- and medium-sized enterprises respond to responsible investment. *Business Strategy and the Environment, 31*(4), 1–14. https://doi.org/10.1002/bse.2981

Ward J. (2002). The role of the board in family business strategy. *Family Business Know-How*, June 2002.

Whiterell B. (2002). Corporate governance and responsibility. *OECD Observer*, no. 234, October 2022.

Yoshikawa, T., Nippa, M., & Chua, G. (2021). Global shift towards stakeholder-oriented corporate governance? Evidence from the scholarly literature and future research opportunities. *Multinational Business Review, 2021*(29), 321–347.

Zhang, Y., & Wei, F. (2021). SMEs' charismatic leadership, product life cycle, environmental performance, and financial performance: A mediated moderation model. *Journal of Cleaner Production, 2021*(306), 127147.

Zhang, L., Zhang, X., An, J., Zhang, W., & Yao, J. (2022a). Examining the role of stakeholder-oriented corporate governance in achieving sustainable development: Evidence from the SME CSR in the context of China. *Sustainability, 2022*(14), 8181.

Zhang, X., Wang, Z., Zhong, X., Yang, S., & Siddik, A. B. (2022b). Do green banking activities improve the banks' environmental performance? The mediating effect of green financing. *Sustainability, 2022*(14), 989.

Zhong, N. (2015). Corporate governance of Chinese privatized firms: Evidence from a survey of non-listed enterprises. *Journal of Comparative Economics, 2015*(43), 1101–1121.

Zumente, I., & Bistrova, J. (2021). Do Baltic investors care about environmental, social and governance (ESG)? *Entrepreneurship and Sustainability Issues, 2021*(8), 349–362.

Zutshi, A., Mendy, J., Sharma, G. D., Thomas, A., & Sarker, T. (2021). From challenges to creativity: Enhancing SMEs' resilience in the context of COVID-19. *Sustainability, 13*(12), 6542. https://doi.org/10.3390/su13126542

Part II

Chapter 7
Best Practice in Sustainable Governance Development Worldwide

7.1 Leading and Lagging Sustainable Governance Elements

Sustainable governance elements are supposed to be developed in harmony, with equal importance of each pillar. However, the notion of sustainable governance is new, and, in practice, company may implement few elements out of all the bunch of the tools or make priorities for one of the elements.

For example, company may develop policies in the fields of ESG, stakeholder communication, anticorruption, etc. But at the same time, there will not be any department responsible for ESG strategy development and implementation, no KPIs, and no compliance system in place. Such extreme case of reduced version of sustainable governance implementation may be called a convenience set of sustainable governance tools: the easiest element to implement is taken onboard, while others are postponed.

Another one option to implement the elements of sustainable governance is by choosing the most financially relevant ones. Such elements are usually connected with the strategy: ESG strategy, ESG risk management, and ESG risk disclosure. They are less regulated, which makes them out of the market and, thus, more significant in the dependence between sustainable governance scores and market value of the company.

These two cases of prioritization are described in the research of Cardoni et al. (2022), where authors presented the real picture of the sustainable governance and its element development on the example of 110 best practice companies. The sample is derived from the GRI database as GRI report is the preliminary benchmark of the leading practice in the field of ESG. Authors took reports from publicly listed companies from seven countries representing all continents: USA, GB, Italy, Russia, Brazil, Australia, and South Africa.

The companies from the sample a large with mean market value of USD3320 million and total assets of USD9790 million. However, companies are extremely

A. Cardoni, E. Kiseleva, *Sustainable Governance*, CSR, Sustainability, Ethics & Governance, https://doi.org/10.1007/978-3-031-37492-0_7

Table 7.1 Descriptive statistics of sustainable governance score and its components

Variable	Obs	Mean	Std. dev.	Min	Max
Market value (USD million)	110	3320	6460	0.2	48,900
Total assets (USD million)	109	9790	29,500	0.5	195,000
Net debt (USD million)	107	1660	5310	−5780	40,900
Revenue (USD million)	109	2020	3640	0.2	27,800
Net income (USD million)	109	1.9	3.8	−1.3	2050
EBITDA (USD million)	108	4.09	6.6	0.04	3590

Source: constructed by the authors

heterogeneous in terms of size and profitability considering standard deviations of the values. The net income after tax has both negative and positive values, which shows that best ESG practice may appear not only for companies with extreme margins. The same applies to the level of debt: companies have wide interval of possible values (Table 7.1).

The methodology of sustainable governance evaluation was similar to the one described in the following book. The primary sources of corporate information used were annual reports and sustainability reports, which are popular tools used by companies to share information on sustainability risks (economic, social, and environmental) and sustainability performance with stakeholders (Du et al., 2017; Dumay & Hossain, 2019). The reports of the sample companies were published in 2018–2020 with reporting period of 2018, in English language.

Concerning sustainability governance score, it was analyzed using a division into three sections:

- Strategy (X1).
- Stakeholder governance body (X2).
- Anticorruption (X3).

Transparency issue was integrated into each of the sections.

Each section could have gained values from 0 to 3 points, where 0 is the absence of information or absence of ESG activities while 3 is the best practice reporting and best practice ESG actions.

The mean values of *X1* and *X2* appeared to be quite close to maximum (2.15 and 2.52, respectively, out of 3), while *X3* mean value is lower (1.75 out of 3). Such difference in mean values can be explained by the high regulation of the corporate governance concerning strategy and board composition, while corruption disclosure and anticorruption initiatives are mostly voluntary in the majority of countries and hence less developed.

The mean total sustainable governance score (*X4*) was evaluated as 6.43 out of 9. Sustainable governance score minimum is over 4.2, which is more than the sum of minimums of its sections. It shows that companies manage the three components of sustainable governance in a balance: if some component has poor performance, it is compensated by the other. At the same time, there is no maximum (9 out of 9) sustainable governance score (*X4*) in a sample (Table 7.2, Appendix A).

Table 7.2 Descriptive statistics of sustainable governance score and its components

Variable	Obs	Mean	Std. dev.	Min	Max
X1 (strategy)	110	2.157	0.599	0.750	3.000
X2 (stakeholder governance)	110	2.522	0.409	1.250	3.000
X3 (anticorruption)	110	1.750	0.497	0.500	2.667
X4 (total sustainable governance as a sum of X1, X2, and X3 for each company)	110	6.433	0.908	4.167	8.333

Source: constructed by the authors

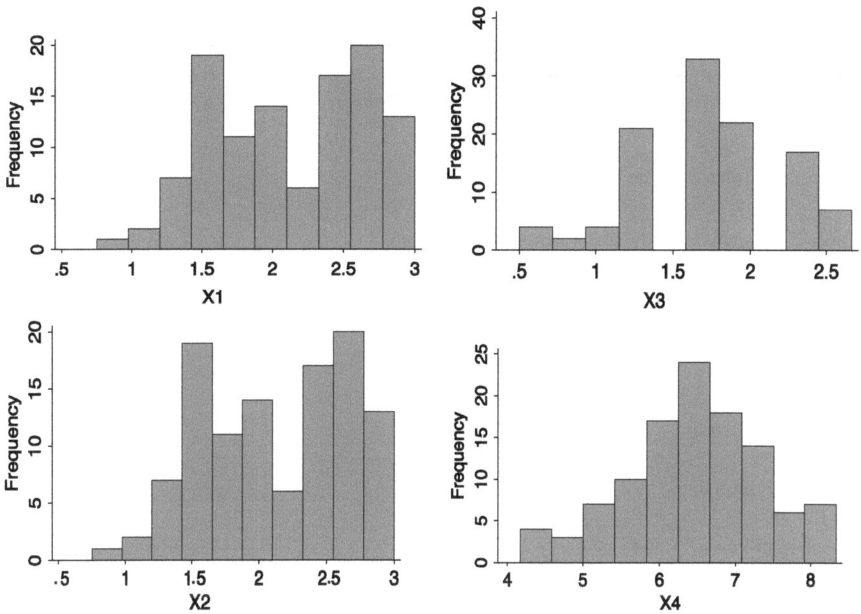

Fig. 7.1 Distribution of sustainable governance score and its components. Source: constructed by the authors

The distribution of the scores is showing that the common widespread practice is formed for the anticorruption block of sustainable governance that is highly regulated in any jurisdiction. In contrast, nonregulated strategic approach and stakeholder governance body metrics have results that are more heterogeneous. In particular, *X4 (total score)* and *X3 (anticorruption)* are close to normal distribution, while *X2 (stakeholder governance)* is left-skewed and *X1 (strategy)* has two peaks (Fig. 7.1).

Breakdown by country (Table 7.3) shows that companies from developed economies, mainly US companies, publish the majority of the best practice reports under analysis. The criteria for choosing a report were the availability of the sustainability report prepared in accordance to GRI standards. Companies from more developed institutional frameworks have more pressure for presenting the nonfinancial information, while stakeholders of such companies are expecting the high level of

Table 7.3 Breakdown by country

	N. of companies	Mean values of sustainable governance score			
		X1-strategy	X2-stakeholder	X3-anti-corruption	X4-total sustainable governance
Developed countries					
USA	71	2.10	2.56	1.61	6.29
Italy	12	2.17	2.56	1.96	6.69
UK	10	2.25	2.56	1.90	6.71
Australia	2	2	2.75	2.33	7.08
Total mean	95	2.12	2.57	1.70	6.40
Emerging countries					
Brazil	5	2.25	2.30	2.13	6.68
Russia	9	2.36	2.12	2.00	6.48
South Africa	1	3.00	3.00	2.00	8.00
Total mean	15	2.37	2.24	2.04	6.65

Source: constructed by the authors

Table 7.4 Breakdown by size in terms of total assets

Total assets, USD	No. of companies	Mean values of sustainable governance score			
		X1-strategy	X2-stakeholder	X3-anti-corruption	X4-total sustainable governance
From 0.5 billion to 7.2 billion	28	2.13	2.54	1.69	6.36
From 7.2 billion to 16.2 billion	27	2.14	2.48	1.91	6.53
From 16 billion to 44 billion	27	2.06	2.48	1.67	6.22
From 44 billion to 1.95 trillion	28	2.29	2.58	1.73	6.62
Total sample	109	2.16	2.52	1.74	6.43

Source: constructed by the authors

transparency and a particular set of quality characteristics. One of such characteristics may be an adherence to GRI standards, which leads to a greater number of GRI reports from developed economies. Apart from this, the formal criteria for reports were the availability of the report in English. Thus, the English-speaking countries appeared to be in priority (USA and UK).

All companies under analysis are large corporations according to EU classification. However, the difference in size between companies is enormous. The quartile analysis of the total assets and market value (Tables 7.4 and 7.5) depicts the scores of sustainable governance development for corporations of different scales of operations. The variation in the levels of sustainable governance may be visible between SMEs and medium size and large corporations; however, large businesses representing best practice do not show extreme fluctuations in values. To this end,

Table 7.5 Breakdown by size in terms of market value

Market value, USD	No. of companies	Mean values of sustainable governance score			
		X1-strategy	X2-stakeholder	X3-anti-corruption	X4-total sustainable governance
From 0.7 billion to 4.5 billion	28	2.13	2.35	1.72	6.20
From 4.5 billion to 12.8 trillion	27	2.20	2.57	1.78	6.56
From 1.3 trillion to 32.6 trillion	27	2.09	2.57	1.72	6.38
From 32.6 trillion to 489 trillion	28	2.20	2.60	1.78	6.60
Total sample	110	2.16	2.52	1.75	6.43

Source: constructed by the authors

Table 7.6 Breakdown by income

Net income, USD	No. of companies	Mean values of sustainable governance score			
		X1-strategy	X2-stakeholder	X3-anti-corruption	X4-total sustainable governance
From negative values to 116 million	28	2.25	2.49	1.67	6.40
From 116 million to 483 million	27	2.18	2.48	1.96	6.61
From 483 million to 1.9 billion	27	1.98	2.54	1.62	6.16
From 1.9 billion to 21 billion	28	2.21	2.58	1.76	6.55
Total sample	109	2.16	2.52	1.74	6.43

Source: constructed by the authors

the best practice for large companies is defined and may be described by a particular set of the common sustainable governance elements.

The age-old riddle about chicken and egg, or ESG performance and financial performance, is a highly arguable topic. Some state that ESG, together with other factors, may enable financial outperformance. At the same time, others claim that only excessive returns may allow such luxury as ESG activities. Large corporations with best practice ESG performance puzzle both of the counterparts since the statistics is not showing a clear answer. Sample companies have varying levels of incomes from negative or close to zero (25%) to billions of net income. However, all of these companies present high results in sustainable governance evaluation (Table 7.6).

The evaluation of each criterion inside the sections of the corporate governance revealed that companies tend to use convenience set of sustainable governance tools. The most popular criteria among companies were the ones that show the following:

- Basic transparency of the issue (short description of the board members, financial risks, anticorruption measures).
- Corporate policies in place or regulatory norms of corporate governance (anticorruption policy, ethical policy, code of conduct, and others).
- Obligatory data (financial risks).
- Easy to implement tools (presentation of the anticorruption policies for new employees).

7.2 Balanced Sustainable Governance Integration

Some researchers presented the models for ESG integration into corporate structures.

Asif et al. (2011) present an integrated management systems approach to corporate sustainability (Asif et al., 2011). According to this model, the process of integrating sustainability includes the identification of key stakeholders and their requirements, prioritization of them, deploying a number of different management systems (usually, ISO standards), and, finally, the integrated management review to track the effectiveness of the system. The model is based on the meta-management of management system integration developed by Asif et al. in 2010. Authors explain the integration of different management systems through an "umbrella" meta-management approach that unites various subsystems at a higher level of abstraction. The use of meta-management is of paramount importance as far as it is impossible for management system to be implemented separately without having an influence on each other. Apart from the meta-management approach, the model of Asif et al. (2011) is designed around the Plan-Do-Check-Act cycle (Deming & Cahill, 2018) and includes all the levels of implementation: strategic, tactical, and operational ones. Moreover, it is based on the involvement of stakeholders in the business. Such inclusion corresponds to the stakeholder theory of company's ESG sustainability (Freeman & McVea, 2001).

Another model, proposed by Eccles et al. (2012), describes three stages of integration: reframing the company's identity, codifying the new identity, and building a supportive organizational culture (Eccles et al., 2012). Reframing the company's identity means leadership commitment and external engagement to ESG topics. The second stage concerns codifying the new identity which involves building internal support for the new identity through employee engagement and mechanisms for execution. A supportive organizational culture, as a last element of the integration model, is supposed to strengthen the results of ESG implementation. The model of Eccles et al. (2012) involves all the stages of analysis (from planning to measurement and evaluation) as well as all the levels of the company (strategic, tactical, and operational). This structure is not presented in a very clear way as it is in Asif et al. (2011) model; however, it goes in more detail on what does it mean to implement each step. More importantly, similar to Asif et al. (2011) model, it also involves stakeholders; however, in a passive form, stakeholders are informed and influenced by the company, but not necessarily involved in a decision-making process.

Baumgartner and Ebner (2010) provide a strategic approach to ESG (Baumgartner & Ebner, 2010). In this paper, they discuss the relation between corporate sustainability strategies, corporate competitive strategies, and sustainability aspects. The authors specified beginning, elementary, satisfying, and sophisticated (outstanding) levels of sustainability in the company. Also they provided the variety of the ESG aspects and described determination of their maturities (for each level of sustainability). The authors applied four types of strategies (introverted, extroverted, and conservative) for each ESG aspect at each stage of maturity. The findings tell that each maturity level needs different types of strategies to implement. Although the authors went further than previous ones and described the contents of the aspects, their levels of maturity, and connection to the strategy type, the results are still too broad. The strategy types describe generic possibilities to deal with the challenge of sustainability but do not describe managerial steps in achieving these strategies.

The attempt to create the theoretical framework that can be more suitable to managerial purposes was undertaken by Whitelock (2019). The model is titled "multidimensional environmental social governance (ESG) sustainability framework." The model is implemented using a purchasing, operations, and supply chain management context and includes five constructs. Two of them make the link to managerial reality; in particular, these constructs describe the distinct management process (e.g., plan, source, make, deliver, and return) and synchronous management (holistic management philosophy when every action of all the levels of the company is focused on the common global goal of the organization). Moreover, the author makes an emphasis on the need of integrated approach: sustainability efforts must be integrated internally, and vertically, within a firm and between departments, plants, and divisions of a firm (Whitelock, 2019).

For the purposes of building a more practical-oriented and precise framework, we modified the Asif et al. (2010, 2011) theories and applied to one of the sustainable governance issues (anti-corruption) (Cardoni et al., 2020), which allowed to build the framework adaptable for integration of any sustainable governance and other ESG issues (Fig. 7.2).

Integrated meta-management framework is based on a plan-do-check-act logic (Deming & Cahill, 2018), which highlights the continuous improvement process imbedded in the model. The framework is integrated throughout all the company, vertically and horizontally, with the consideration of other E, S, and G issues of the business. The framework is based on the meta-management approach—the overarching frame for different management disciplines (e.g., strategy management, risk management) that allows the transformation process to be effective (Uhl & Gollenia, 2016).

In the *"Plan" phase*, the first step is that of *stakeholder requirements definition*. Various stakeholder requirements should be taken into account. Multi-stakeholder initiatives comprised of representatives from government, private sector, and civil society organizations help in finding material topics out of E, S, and G issues that are

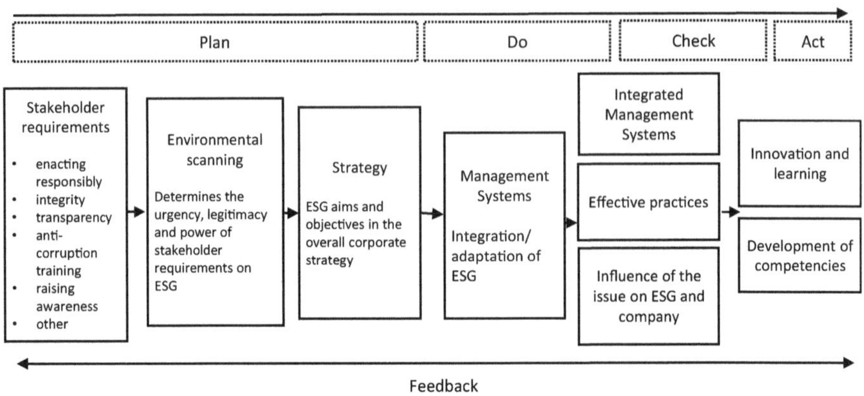

Fig. 7.2 Integrated meta-management framework. Source: constructed by the authors basing on Asif et al. (2010, 2011)

vital nowadays. Such stakeholders may serve company as navigating organizations in the implementation of the following step. For example, one of the most popular multi-stakeholder initiative that aggregates the interests of stakeholders is the International Organization for Standardization (ISO) that brings together experts to share knowledge and develop voluntary, consensus-based, market relevant international standards that provide solutions to sustainability challenge.

The environmental scanning is of paramount importance in the plan stage as it is aimed at defining the urgency, legitimacy, and power of the stakeholders' expectations with reference to corporate ESG. Environmental scanning helps in identifying key stakeholders and prioritizing them within a context of constrained resources. Each company has its own unique stakeholder surrounding (Deegan & Shelly, 2014) and particular resource availability, which makes the results of the environmental scanning special for each enterprise. As a result, companies form the understanding of the surroundings and develop the appropriate context that serves as a reasonable and credible guide for action (Choo, 1996).

After the environmental scanning is undertaken, *ESG strategy* is derived with the aim to meet the identified stakeholder requirements. Once the goals, mission, and vision are set, focus is shifted from the strategic level toward the tactical and operational level (Asif et al., 2010).

Important 7.1 The step of strategy implementation is vital: strategic ESG allows to gain the competitive advantage, while integration of standardized ESG-related management systems allows to achieve the mere survival level of the ESG activity (Ioannou & Serafeim, 2019; Porter & Kramer, 2006).

The "Do" step starts with the *integration or development of the management systems* that allows to implement the sustainable governance aims and objectives from the overall ESG strategy. With this aim, the model of Asif et al. (2011) proposes to use ISO standards. The choice of ISO standards or other guidance is explained by the fact that standardization simplifies the process of ESG management and helps in defining the scopes of the topic: company needs only to comply with the requirements from the standards. Thus, the management system becomes a simple-to-use structure of managing processes (Schwartz & Tilling, 2009). However, the standards may isolate the company from real stakeholder requirements and company's strategy (Schwartz & Tilling, 2009). Moreover, standards are representing compliance or "checkoff" approach that is proven to be ineffective. Thus, the standards implementation was eliminated from the following framework, highlighting the role of non-standardized approach. Instead, the example of best practice management system from the case study is provided in this chapter.

The "Check" step includes the estimation of *success of management practices* as well as the effectiveness of meta-management integration throughout all the organization and for E, S, and G issues. In order to undertake the check step, the key performance indicators from ESG strategy should be analyzed. *The level of integration* can be analyzed through a number of fundamental changes at the strategic, tactical, and operational levels: corporate governance and strategy, integrated objectives, integrated manuals, integrated procedures, structures (integrated functions, audits, and trainings), integrated processes, continuous improvement infrastructure, routines, and integrated reviews. As far as integration with ESG and other spheres is concerned, it should be verified *what effect ESG activities have on ESG and other processes of the company.*

The "Act" step implies that evaluation results are to be processed and used for *future development* of the meta-management system. New knowledge needs to be integrated explicitly (by manuals, procedures, databases, work instructions, and other key documents) and tacitly (employees' experiences and skills) (Asif et al., 2011).

The cycle is ongoing and has an ongoing communication with stakeholders for reporting on the results of ESG activities as well as for receiving feedback from them.

Being developed based on one sustainable governance issue (anti-corruption) and with the verification on a real company, framework may be implemented for integrating sustainable governance elements as well as any ESG elements in the company. The PDCA cycle, familiar to any manager, allows using framework as a practical tool in the corporate plans and road maps for ESG sphere development. The next part of the chapter presents the key success factors for implementing the framework effectively and leading company to the financial outperformance.

7.3 Success Factors in Sustainable Governance Integration

While implementing the framework, there may be a question on what the priorities in the process are and how to make an implementation successful. The case of an Italian manufacturing company in this chapter presents the answer to this question on the example of effective integration of anti-corruption issue (Cardoni et al., 2020).

Italian manufacturing company Acciai Speciali Terni Spa (AST) is an Italian company with about 2400 employees and 1.7 billion EUR turnover as of 2019. The core business consists of flat rolled stainless steel products, for which AST is the market leader in Italy and one of the top 4 manufacturers in Europe. In addition to flat rolled steel products, the company produces electro welded stainless tubes and forgings made of special steel. The headquarter is located in Germany. Parent company, Thyssenkrupp, consists of 159,426 employees in 78 countries and has turnover of 21.9 billion euros (as of 2019 as well). Thyssenkrupp business operations are organized in five business areas: components technology, elevator technology, industrial solutions, materials services, and the discontinued operation of Steel Europe (Cardoni et al., 2020).

AST is located in the sensible environment in relation to anti-corruption. The rate of corruption in Italy is 52/100 which is relevant to 53rd position out of 180 countries ("Corruption Perceptions Index 2018," 2019). The level of corruption is higher than in many other European countries. In order to improve the anti-corruption climate, Italian government is motivating companies for active participation in anticorruption movement. Government is promoting new multi-shareholder approach to anticorruption, according to which companies are called to be paramount participants in combating corruption. Moreover, the manufacturing sector that the company belongs to is one of the five most corrupted industries (Kottasova, 2014).

The company is stimulated to build the workable anti-corruption system in order to prevent corruption cases. Indeed, the anti-corruption management of AST represents the practice that goes far beyond the mere compliance to government requirements. According to Italian legislation, companies only need to adopt a simple compliance program, while AST not only has a compliance program but also builds a more elaborate and diverse system of anti-corruption.

The experience of AST in the field of anti-corruption implementation shows the critical factors that differ them from traditional approach to anti-corruption and from other organizations. These factors are supposed to be the main pillars in implementing the framework.

These key success factors, derived from the interviews with AST management and theoretical verifications, are presented for each step in the framework (Fig. 7.3).

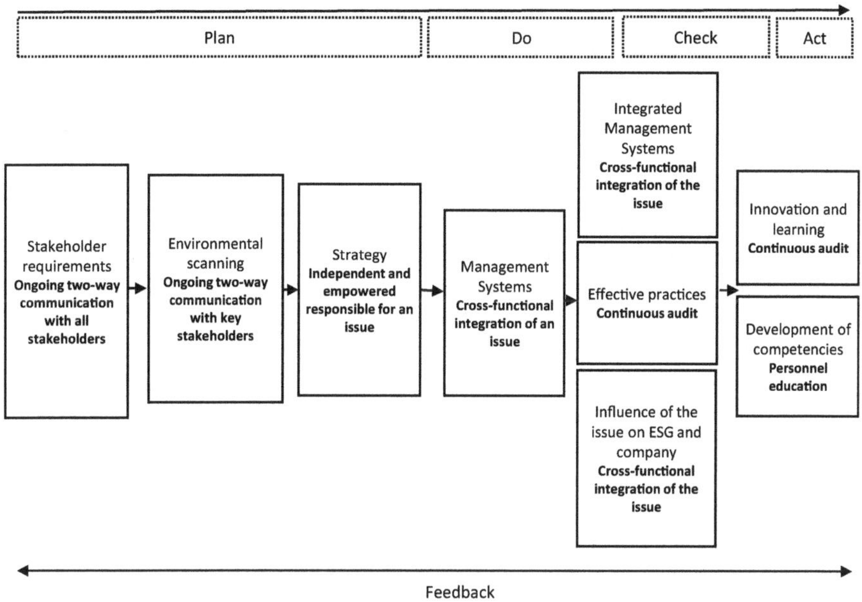

Fig. 7.3 Key success factors for framework implementation. Source: constructed by the authors basing on Asif et al. (2010, 2011)

7.3.1 Ongoing Two-Way Communication with Stakeholders

The expectations of internal stakeholders (process owners in functions/departments, employees, CEO) and external stakeholders (anti-corruption initiatives, government, suppliers, clients, local community, and so on) are a relevant issue. Such involvement has the potential to provide benefits and correlates with the theoretical concept of sustainable governance (Lombardi et al., 2019), in which stakeholder communication plays the main role. External engagement makes the company's mechanisms of execution stronger since stakeholder pressure leads to the constant improvement in its quality (Eccles et al., 2012).

Whitelock (2019) also stated that expanding the level of integration externally can enable a company to continuously learn from its partners and gain enhanced efficiency and efficacy benefits (Whitelock, 2019). Moreover, organizations are directed to understand the ESG issue in conversation with others while exchanging ideas with them in an iterative sense-giving (environmental scanning) and sense-making process (feedback to stakeholders) (Gioia & Chittipeddi, 1991). Finally, stakeholder expectations regarding ESG are an ever-changing topic and are regarded carefully and on a frequent basis.

The systematic stakeholder engagement that AST has undertaken results in increased accountability to a range of stakeholders. Accountability, in turn, strengthens trust between the organization and its stakeholders.

7.3.2 Independent and Empowered Supervisor for Anti-corruption

Another success factor is connected to the level of the ESG issue adoption in the company (Lombardi et al., 2019). The position of responsible manager supervisor in governance structure is a relevant issue. It is necessary to appoint a person in charge for an issue (in relation to anti-corruption, it is usually an Internal Control System Director), positioned at the same level of the other process owners such as the CFO, General Manager, Sales Director, etc. This supervisor needs to be autonomous and independent and accountable at least to the CEO or the Board of Directors, and decisions about their designation/discharge should be made collectively.

Such hierarchy will allow the supervisor for an issue to directly participate in the strategic decision-making of the company and put the issue into the corporate strategy agenda. High position of the ESG issue supervisor may guarantee the integration of ESG objectives into corporate strategy, while the low level of the supervisor will not leave a chance for strategic approach to be realized.

Moreover, the supervisor will act as a pivotal figure and facilitator for all actors involved in ESG activities. The right level will enable operational synergies among process owners, and the common language will be built. Moreover, the high positioning of the supervisor will help to avoid the duplication of roles, responsibilities, and activities.

The independence of the supervisor will help gain objectivity in actions, which is vital for the topic of ESG and anticorruption in particular.

> You have to clearly identify a unique supervisor for anticorruption, who is in charge of enabling the synergies between all the key anticorruption stakeholders—Governance Director and Business Continuity Officer.

To overcome detrimental checkoff approach, the empowerment of an ESG issue (s) supervisor is needed. In case of anti-corruption, an organization may establish an anti-corruption committee consisting of senior vice presidents for marketing and sales, auditing, operations, human resources, and other key officers. The same may be applied to other issues: the environmental/human resources/ESG/sustainability or any other committee may take place in the organization depending on the priorities of a company and its stakeholder environment.

7.3.3 Cross-Functional Integration of an Issue

The cross-functional collaboration in the fight against corruption or ESG development is of paramount importance. Process owners (marketing department, CFO, production department, etc.) are no longer the unique parties responsible for the risks affecting their areas because today world complexity makes it difficult for process owners to identify all the ESG risks. That is why companies must enable synergies between process owners and the ESG issue supervisor. Together they can define the key risks and response strategies. The supervisor establishes the best types of control tools for monitoring the risks in the process owner field and shares the risks with the process owners; they become not only a controller but also a co-owner of the risks.

The meta-management approach, which is used for anti-corruption integration at AST company, considers the whole system, rather than ad hoc quick fixes that solve only a part of the problem (Asif et al., 2010; Stiles & Uhl, 2012). Moreover, anti-corruption as a part of culture (Erp et al., 2015) or CSR also creates a challenge in implementation across all whole company, leading to the need for cross-functional integration.

7.3.4 Continuous Auditing

In the modern world of big data, the methods of external auditors, who come twice per year and take a sample of data, are completely obsolete. One hundred percent of data should be analyzed in order to make it impossible to avoid the check. Thus, control must become smart and digital and utilize new technologies to ensure continuous monitoring instead of "ex ante" approvals. Continuous auditing allows not only to analyze 100% of the data but also helps to concentrate on which high risk issues need to be worked on, minimizing the human resources.

Continuous auditing is a potential successor to the traditional auditing paradigm. It is continuous, proactive, and automated and uses whole data population unlike the traditional one (Chan & Vasarhelyi, 2011; Groomer & Murthy, 2018).

Most importantly, ongoing monitoring helps to develop double-loop learning, which will question the underlying assumptions and governing principles of the current model, thus leading to its modification (Argyris, 1977). The learning process is imperative for continuous improvement and necessary for the long-term survival of businesses. Organizational learning, in particular double-loop learning, is an essential part of the meta-management approach, thus a critical factor in ESG implementation.

7.3.5 Personnel Education

The professional profile of the ESG issue supervisor should be multi-competent in guaranteeing sustainable governance and CSR (Welford, 2007). In the checkoff approach, people in charge usually come from the legal sphere, equipped with the obligatory requirements and who are obedient to the law. However, in the context of the integrated approach to ESG, the supervisor needs to manage comprehensive situations and have the ability to navigate multitasking teams.

Personnel competencies, especially ESG supervisor competencies, are of paramount importance as people are crucial for meta-management implementation

(Stiles & Uhl, 2012). Meta-management includes cross-functional communication and collaboration with internal and external stakeholders, which leads to the need for multitasking teams with a multi-competent leader.

Appendix

Table 7.7 Sustainable governance scores of the companies

Company name	X1 (strategy)	X2 (stakeholder governance)	X3 (anti-corruption)	X4 = X1 + X2 + X3
Fiat Chrysler Automobiles	3.000	3.000	2.333	8.333
Snam	3.000	2.500	2.667	8.167
ConvaTec	2.750	3.000	2.333	8.083
Eni S.P.A.	3.000	3.000	2.000	8.000
Gold Fields	3.000	3.000	2.000	8.000
Hess Corporation	3.000	3.000	2.000	8.000
Polyus	2.500	2.750	2.667	7.917
General Motors Company	2.500	3.000	2.333	7.833
Varex Imaging	2.750	2.750	2.307	7.807
Saipem	2.750	3.000	2.000	7.750
Woodside	2.500	2.500	2.667	7.667
Jones Lang LaSalle	2.750	2.750	2.000	7.500
Cielo	3.000	2.167	2.333	7.500
Alcoa Corporation	3.000	2.750	1.667	7.417
Compass Minerals	2.750	3.000	1.667	7.417
DS Smith	2.750	3.000	1.667	7.417
ON Semiconductor	2.000	2.750	2.667	7.417
BP	2.000	3.000	2.333	7.333
Pearson	2.500	2.500	2.333	7.333
Exelon corporation	2.750	2.750	1.667	7.167
LSR group	2.250	2.250	2.667	7.167
Mettler Toledo	2.750	2.750	1.667	7.167
Novatek	3.000	2.500	1.667	7.167
Prudential financial. Inc.	2.500	3.000	1.667	7.167
Synthomer	3.000	2.500	1.667	7.167
Xcel energy	2.500	3.000	1.667	7.167
Lockheed Martin Corporation	2.750	2.000	2.333	7.083
Crestwood Equity Partners LP	3.000	2.750	1.332	7.082

(continued)

Table 7.7 (continued)

Company name	X1 (strategy)	X2 (stakeholder governance)	X3 (anti-corruption)	X4 = X1 + X2 + X3
Itaúsa	2.000	3.000	2.000	7.000
MetLife	2.500	2.500	2.000	7.000
Xylem	2.750	2.250	2.000	7.000
Citi	2.500	2.750	1.667	6.917
Freeport-McMoRan, Inc.	3.000	2.250	1.667	6.917
International Paper	2.250	3.000	1.667	6.917
Polymetal International	2.750	2.500	1.667	6.917
Banco do Brasil	3.000	1.500	2.333	6.833
Mondi	1.500	3.000	2.333	6.833
The Timken Company	2.500	2.500	1.833	6.833
Mediaset Group	2.000	2.750	2.000	6.750
Mylan	2.000	2.750	2.000	6.750
BRF S.A.	2.000	2.417	2.333	6.750
AptarGroup	1.750	2.250	2.667	6.667
HCP. Inc.	2.750	2.250	1.667	6.667
PVH Corporation	2.750	2.250	1.667	6.667
The Hartford	2.500	2.500	1.667	6.667
Wells Fargo & company	2.000	3.000	1.666	6.666
Hammerson	2.250	3.000	1.333	6.583
Humana	1.500	2.750	2.333	6.583
Moody's Corporation	2.750	2.500	1.333	6.583
PPL Corporation	2.750	2.500	1.333	6.583
Prysmian Group	2.750	1.500	2.333	6.583
State Street Corporation	2.500	2.750	1.333	6.583
Merck & co., Inc.	2.000	2.500	2.000	6.500
OceanaGold	1.500	3.000	2.000	6.500
Regency Centers	1.750	2.750	2.000	6.500
Unicredit	1.500	3.000	2.000	6.500
Cushman & Wakefield	2.500	2.250	1.667	6.417
Rosseti	2.500	2.250	1.667	6.417
Sempra Energy	2.000	2.750	1.667	6.417
Financial Corporation Sistema	2.250	1.500	2.667	6.417
American Electric Power	2.500	2.500	1.333	6.333
ConocoPhillips	2.500	2.500	1.333	6.333
En+ Group	2.750	2.250	1.333	6.333
Ferrari	2.000	3.000	1.333	6.333
Fluor	1.750	2.250	2.333	6.333
Kimberly-Clark	2.250	2.750	1.333	6.333

(continued)

Table 7.7 (continued)

Company name	X1 (strategy)	X2 (stakeholder governance)	X3 (anti-corruption)	X4 = X1 + X2 + X3
Northrop Grumman	1.750	2.250	2.333	6.333
Flex	2.750	2.250	1.327	6.327
Transneft	3.000	1.917	1.333	6.250
Comerica Bank	2.000	2.500	1.667	6.167
El.En. S.p.A.	2.250	1.917	2.000	6.167
Iron Mountain	1.500	3.000	1.667	6.167
3M	1.250	2.500	2.333	6.083
Bunge	1.750	3.000	1.333	6.083
Caesars Entertainment	2.500	2.750	0.833	6.083
Balchem	2.750	2.250	1.000	6.000
Caterpillar	2.750	2.750	0.500	6.000
Advanced Micro Devices	1.500	2.750	1.667	5.917
Arconic	1.500	2.750	1.667	5.917
Derwent London	1.750	2.500	1.667	5.917
KELLY Services	2.000	2.750	1.167	5.917
Prologis	1.750	2.500	1.667	5.917
Salini Impregilo	1.750	2.500	1.667	5.917
Endeavour Mining	2.500	1.333	2.000	5.833
Ormat Technologies. Inc.	1.750	2.750	1.333	5.833
Alacer Gold	1.500	2.667	1.667	5.833
CBRE	1.000	2.750	2.000	5.750
Air Products	1.500	2.500	1.667	5.667
Eversource Energy	1.750	2.250	1.667	5.667
FMC Corporation	1.250	2.750	1.667	5.667
EQT corporation	2.000	2.250	1.333	5.583
Mastercard	1.500	2.750	1.333	5.583
DTE Energy Company	1.500	3.000	1.000	5.500
GEDI Spa	1.250	2.417	1.833	5.500
Kansas City Southern	1.750	1.750	2.000	5.500
XPO logistics. Inc.	1.500	2.000	2.000	5.500
Cabot Corporation	1.500	2.500	1.333	5.333
Gazprom	1.500	2.167	1.667	5.333
JBS S.A.	1.250	2.417	1.667	5.333
Rostelecom	1.500	1.500	2.333	5.333
International Game Technology	1.250	1.750	2.000	5.000
PPG Industries	1.250	2.750	1.000	5.000
Principal Financial Group	1.500	2.167	1.333	5.000
Archer Daniels Midland	1.000	2.750	1.167	4.917

(continued)

Table 7.7 (continued)

Company name	X1 (strategy)	X2 (stakeholder governance)	X3 (anti-corruption)	X4 = X1 + X2 + X3
Cree	1.500	2.500	0.833	4.833
Brookfield Properties —US Retail	1.250	3.000	0.500	4.750
Ingevity	1.500	2.250	0.667	4.417
Arthur J. Gallagher & Co.	2.000	1.250	1.000	4.250
Italmobiliare Investment Holding	0.750	2.167	1.333	4.250
Vornado	1.500	2.167	0.500	4.167

Source: constructed by the authors

References

Argyris, C. (1977, September 1). Double loop learning in organizations. *Harvard Business Review*. https://hbr.org/1977/09/double-loop-learning-in-organizations

Asif, M., Joost de Bruijn, E., Fisscher, O. A. M., & Searcy, C. (2010). Meta-management of integration of management systems. *The TQM Journal, 22*(6), 570–582. https://doi.org/10.1108/17542731011085285

Asif, M., Searcy, C., Zutshi, A., & Ahmad, N. (2011). An integrated management systems approach to corporate sustainability. *European Business Review, 23*(4), 353–367. https://doi.org/10.1108/09555341111145744

Baumgartner, R. J., & Ebner, D. (2010). Corporate sustainability strategies: Sustainability profiles and maturity levels. *Sustainable Development, 18*(2), 76–89. https://doi.org/10.1002/sd.447

Cardoni, A., Kiseleva, E., Arduini, S., & Terzani, S. (2022). From sustainable value to shareholder value: The impact of sustainable governance and anti-corruption programs on market valuation. *Business Strategy and the Environment*. https://doi.org/10.1002/bse.3328

Choo, C. W. (1996). The knowing organization: How organizations use information to construct meaning, create knowledge and make decisions. *International Journal of Information Management, 16*(5), 329–340. https://doi.org/10.1016/0268-4012(96)00020-5

Cardoni, A., Kiseleva, E., & Lombardi, R. (2020). A sustainable governance model to prevent corporate corruption: Integrating anticorruption practices, corporate strategy and business processes. *Business Strategy and the Environment, 29*(3), 1173–1185. https://doi.org/10.1002/bse.2424

Chan, D. Y., & Vasarhelyi, M. A. (2011). Innovation and practice of continuous auditing. *International Journal of Accounting Information Systems, 12*(2), 152–160. https://doi.org/10.1016/j.accinf.2011.01.001

Corruption Perceptions Index 2018. (2019). www.transparency.org. Retrieved March 27, 2019, from https://www.transparency.org/cpi2018

Deegan, C., & Shelly, M. (2014). Corporate social responsibilities: Alternative perspectives about the need to legislate. *Journal of Business Ethics, 121*(4), 499–526. https://doi.org/10.1007/s10551-013-1730-2

Deming, W. E., & Cahill, K. E. (2018). *The new economics for industry, government, education* (3rd ed.). The MIT Press.

Du, S., Yu, K., Bhattacharya, C. B., & Sen, S. (2017). The business case for sustainability reporting: Evidence from stock market reactions. *Journal of Public Policy & Marketing, 36*(2), 313–330. https://doi.org/10.1509/jppm.16.112

Dumay, J., & Hossain, M. A. (2019). Sustainability risk disclosure practices of listed companies in Australia. *Australian Accounting Review, 29*(2), 343–359. https://doi.org/10.1111/auar.12240

Eccles, R. G., Perkins, K. M., & Serafeim, G. (2012). How to become a sustainable company. *MIT Sloan Management Review, 53*(4), 43-+.

Erp, J. van, Huisman, W., & Walle, G. V. (2015). *The Routledge Handbook of White-Collar and Corporate Crime in Europe*. Routledge.

Freeman, R. E., & McVea, J. (2001). *A stakeholder approach to strategic management* (SSRN scholarly paper ID 263511). Social Science Research Network. https://papers.ssrn.com/abstract=263511

Gioia, D. A., & Chittipeddi, K. (1991). Sensemaking and sensegiving in strategic change initiation. *Strategic Management Journal, 12*(6), 433–448. https://doi.org/10.1002/smj.4250120604

Groomer, S. M., & Murthy, U. S. (2018). Continuous auditing of database applications: An embedded audit module approach. In Continuous Auditing (world; pp. 105–124). Emerald Publishing Limited. https://doi.org/10.1108/978-1-78743-413-420181005

Ioannou, I., & Serafeim, G. (2019). Corporate sustainability: A strategy? *Harvard Business School,* SSRN Journal. https://doi.org/10.2139/ssrn.3312191

Kottasova, I. (2014, December 2). World's most corrupt industries. *CNNMoney*. https://money.cnn.com/2014/12/02/news/bribery-foreign-corruption/index.html

Lombardi, R., Trequattrini, R., Cuozzo, B., & Cano-Rubio, M. (2019). Corporate corruption prevention, sustainable governance and legislation: First exploratory evidence from the Italian scenario. *Journal of Cleaner Production, 217*, 666–675. https://doi.org/10.1016/j.jclepro.2019.01.214

Porter, M. E., & Kramer, M. R. (2006). Strategy and society: The link between competitive advantage and corporate social responsibility. *Harvard Business Review, 23*(5), sd.2007.05623ead.006. https://doi.org/10.1108/sd.2007.05623ead.006

Schwartz, B., & Tilling, K. (2009). 'ISO-lating' corporate social responsibility in the organizational context: A dissenting interpretation of ISO 26000. *Corporate Social Responsibility and Environmental Management, 16*(5), 289–299. https://doi.org/10.1002/csr.211

Stiles, P., & Uhl, A. (2012). Meta management: connecting the parts of business transformation. *Methodology, 3*, 24–29.

Uhl, A., & Gollenia, L. A. (2016). *A handbook of business transformation management methodology*. Routledge.

Welford, R. (2007). Corporate governance and corporate social responsibility: Issues for Asia. *Corporate Social Responsibility and Environmental Management, 14*(1), 42–51. https://doi.org/10.1002/csr.139

Whitelock, V. G. (2019). Multidimensional environmental social governance sustainability framework: Integration, using a purchasing, operations, and supply chain management context. *Sustainable Development., 27*, 923. https://doi.org/10.1002/sd.1951

Chapter 8
Sustainable Governance in Emerging Economies

Sustainable governance factor is mainly acknowledged as the most powerful factor out of ESG issues, which were discussed in previous chapters. This thesis is especially relevant for the emerging markets (Garanina & Muravyev, 2020; Ting et al., 2020).

Developed economies have corporate governance norms and regulations, pushing companies to the development of sustainable corporate governance. This creates a bulk of companies with the standardized set of corporate governance instruments. Investors and other stakeholders start to perceive such corporate governance as an entrance barrier to the market rather than a reason to pay extra and do not put an attention to the slight changes in governance between companies, simply verifying whether company has standard corporate governance or not.

Companies from developed economies are always in a state of struggle with primary compliance to new laws, while the best practice becomes too expensive and difficult to achieve in such a stringent environment. Corporate governance systems that operate in an already developed governance arena may not provide any additional incentive for the market to value changes in governance in terms of market value. Cornell and Damodaran suggest the same dependence, explaining it by reference to market equilibrium. According to them, there is an adjustment period during which highly rated ESG stocks will outperform low ESG stocks; however, once an equilibrium point is reached, the expected returns offered by the former stocks decline (Cornell & Damodaran, 2020).

Highly developed but highly standardized approach under governmental enforcements in developed economies often leads to a checkoff approach to corporate governance, when company ticks the rules as implemented instead of assessment of the suggested tools. One of the most visible tools in the ESG field that can be approached with a checkoff method is the publication of policies in the areas of ESG, environment, social, anti-corruption, ethics, etc. Such policies address how a company manages and monitors its impact on the environment or stakeholders, both in its own operations and those of its supply chain. Creating a policy involves more than just printing paper; it also requires a series of activities, such as auditing the

A. Cardoni, E. Kiseleva, *Sustainable Governance*, CSR, Sustainability, Ethics & Governance, https://doi.org/10.1007/978-3-031-37492-0_8

corporate ESG system, analyzing best practices and peer benchmarks, communicating with different departments, integrating the policy into the strategy preparation process, and assigning responsible units. These activities ensure the productive implementation of the policy that aligns with corporate strategy, shared by managers and personnel, implemented and controlled for compliance in the company. However, a checkoff approach to policy creation can render the document an irrelevant piece of paper with the analysis of benchmarks as the most valuable result.

Example 8.1 Anti-corruption policy at Gazprom

An example of a formal ESG tool is presented through a company that operates in a less strictly regulated environment of developing economy, which enhances the possibility of the company developing a purposeful policy. Additionally, the company is extremely large, attracting national and international attention and operating in a high-risk environment for corruption (Corruption Perceptions Index 2018, 2019). These factors motivate the company to take tangible action in the fight against corruption.

In particular, the case company is PJSC Gazprom, a Russian majority state-owned multinational energy corporation with total revenues equaling $87.7 billion as of 2020 and a number of employees equaling 479,200 people (Gazprom, 2022).

The Gazprom Anti-corruption Policy aims to establish a transparent and unified approach to fulfilling the preventive anti-corruption obligations of all companies within the group, to the benefit of civil society, shareholders, and employees. The policy reflects the commitment of the management and employees of the group to comply with ethical standards; conduct business in a legitimate, open, and honest manner; enhance the corporate culture; adhere to the best corporate governance practices; and uphold a respectable business reputation.

The policy pursues several objectives. Firstly, it aims to ensure compliance with the applicable anti-corruption laws. Once the minimum normative requirements are met, the company aims to focus on improving its internal anti-corruption regulations. The creation of preventive anti-corruption mechanisms and communication tools is a planned objective to ensure that investors, counterparties, members of management bodies, and employees all share a company-wide anti-corruption consciousness and a common understanding of the company's zero-tolerance policy toward corruption.

Apart from stating aims and objectives, policy declares competences of the preventive anti-corruption management bodies, design principles of a preventive anti-corruption risk management, and internal control system, together with a package of preventive anti-corruption measures.

Out of the measures, Gazprom Policy states the following:

(continued)

Example 8.1 (continued)

- Anti-corruption risk monitoring, identification, and assessment, especially for corruption exposed areas (e.g., procurements).
- Integration of anti-corruption tools across group companies and all departments.
- Tracking and evaluation of changes in anti-corruption laws in corruption-related cases (including international experience).
- Training employees on compliance and updating on changes.
- Encouragement of employees to comply with the policy and actively participating in the process.
- Internal control of the system.

Policy also includes description of anti-corruption standards and their enforcement, as well as the employees' liability for noncompliance with the policy (Tumanov, 2021).

ESG agenda of *developing countries* and sustainable corporate governance in particular are at an earlier stage of its formation; however, the importance and relevance of sustainable development issues is only increasing. Regulators, exchanges, investors, media, and consumers expect businesses to contribute to ensuring a safe and decent future, as well as a transparent approach regarding its impact on the environment and society. At the same time, mandatory requirements are in a less developed state compared to developed markets, which distinguishes best practice companies from their competitors at more extent.

8.1 The Role of Cultural Peculiarities in Sustainable Governance Practice: Russian Case

Corporate governance is highly intertwined with ethical norms and cultural peculiarities, which may be perfectly explained on the example of Russian economy.

Corporate governance in Russia is considered distinct from Anglo-Saxon or continental corporate governance models due to several exclusive characteristics.

Russian companies often have *weak stakeholder involvement* in their governance processes. This is because stakeholders typically lack the necessary resources and capabilities to effectively communicate their requests to the company and often lack experience in collaborative actions. The government is the most influential stakeholder, as it holds significant power over almost all aspects of a company's operations. This includes resource allocation, employee relations, certification for specific activities, environmental standards and penalties for noncompliance, social policies, local community development, etc.

In Russia, there are also fundamental differences related to the *structure of ownership*: instead of dispersed ownership of shares, businesses have a system of key shareholders with controlling stakes, often having government bodies as a controlling shareholder. Due to the high role of the state in the business, corporate mechanisms become similar to public management; hence, corporate governance transforms and becomes less relevant in the company's life, often adopting a formal checkoff approach and sometimes imitating the corporate governance instead of functioning as an efficient system (Ausan, 2021).

The modifications of ESG under the state-oriented economies with paternalistic patterns and roles of the companies may be best interpreted with the use of *CSR-based political legitimacy concept* (Zhao, 2012). In line with this theory, government authorities believe that firms as members of society have a responsibility to make positive contributions to better social and environmental conditions. At the same time, companies undertake sustainability practices to obtain the state legitimacy. Companies generally focus on two aspects: formal policy compliance and the public sector's capacity in social and environmental areas. In other words, companies ensure they comply with environmental and social legislation and assist the state in fulfilling its social obligations by participating in government projects, collaborating with government funding, and even assuming some social obligations on behalf of the state (Zhao, 2012).

Such centralized and standardized approach can enhance the adoption of sustainable development practices, align stakeholder expectations, and simplify navigation of ESG issues for the benefit of society. Additionally, such an approach promotes consistency in the field, addressing a common challenge in ESG, which is the lack of comparability in practices and performance data.

At the same time, under the condition of tight interconnection between business and the state in ESG sphere, the sustainability activities may become a clue to gaining political recognition rather than competitive advantage. Additionally, similar to the experience of developed economies, standardized ESG activities may increase costs for the company without a contribution to a greater economic performance (Cardoni et al., 2020; Ioannou & Serafeim, 2019; Orlitzky, 2015; Porter & Kramer, 2006).

Apart from the high role of government in the ESG agenda, Russian business environment is largely influenced by the transition to a market economy, but at the same time, it is mostly based on the *behavioral norms* from the Soviet Union period keeping the ethical heritage from the tsarist era (Kuznetsov & Kuznetsova, 2003; Mccarthy & Puffer, 2008). The Russian market is sometimes characterized as "chaotic" form of capitalism, dominated by state control, highly corrupted business environment, and the "unrule" of law (Crotty, 2016). However, it is important to note that this criticism stems from attempts to apply the same theoretical constructs used in developed economies to developing markets. Traditional Russian values are often incompatible with a purely capitalistic, market-oriented system and cannot be easily explained using traditional stakeholder, institutional, or agency theories.

The *integrative social contracts theory (ISCT)* by Donaldson and Dunfee (1994, 2002) is applicable for Russian sustainable corporate governance context

(Donaldson & Dunfee, 1994, 2002; Mccarthy & Puffer, 2008) since it adopts the agency theory to the local market by recognizing what the underlying basis is for legitimacy of local norms. Authors highlighted the importance of personal networks in Russian business and examined three traditional Russian cultural dimensions: trust, disrespect for private property, and personal favors. According to Mccarthy and Puffer (2008), these traditional, culturally based local norms are likely to prevail for some time in the decision-making process of the part of Russian managers and board directors (Mccarthy & Puffer, 2008).

Russian ethical norms may bring fruitful results in many fields, but these particular three dimensions derived by Mccarthy and Puffer (2008) are mostly harmful to financial performance of a firm as they may significantly violate the efficiency.

Starting with trust cultural dimension, the tendency to distrust individuals or organizations that fall outside the sphere of personal relationships means no public confidence in the actions of company managers, boards of directors, and such entities as auditors, financial institutions, and government oversight bodies, which, in turn, leads to ineffective relationships with them.

Going forward to the second dimension, the ignorance of private ownership rights of enterprises and little distinction between owners, managers, and other stakeholders lead to disorientation about beneficiaries of business operations and misuse of corporate resources for personal aims.

At last, personal favors dimension is literally equal to corruption in business relationships, which brings direct detrimental losses as well as reputational impacts.

To sum up, the high importance of personal networks means that the final goal of the local ethical behavior is the personal or in-group wealth maximization by the expense of company's value.

Thus, in making sustainable governance effective, *the main role of the sustainable corporate governance* in Russia is twofold. Firstly, it needs to align local norms (trust, disrespect for private property, and personal favors) to internationally accepted hyper norm of law prevalence. Sustainable governance also must broaden the in-group wealth maximization orientation to the more standardized perspective, where the company itself is perceived as such an in-group. In particular, it may build the trust among all parties in the corporate governance system by creating sound rules of the game (corporate governance politics and norms), strong control system for the rules implementation, and sound results of such control. The private property rights ignorance is to be eliminated by increasing transparency and enlarging the distinction between owners, managers, and other stakeholders, where third parties do not see company's assets as their own properties for self-enrichment. Finally, the strong compliance and anti-corruption policies must be in place to avoid the well-spread practice of personal favors.

By implementing sustainable governance tools, a common law can be established that operates for all stakeholders in corporate governance. This will limit the abilities of small groups with their tendency to prioritize in-group utility at the expense of the company's overall value. As a result, corporate communication will become more efficient, leading to improved overall corporate performance and higher financial results for the enterprise.

Cultural peculiarities must always be taken into account when analyzing sustainable corporate governance in companies from developing economies. The term "developing economies" encompasses a variety of different countries, each with its own unique historical background and ethical norms in business practices. This can greatly influence the structure of sustainable corporate governance, as illustrated below through examples of specific companies from China and India.

8.2 Sustainable Corporate Governance Practice

8.2.1 Chinese Case

China has the greatest manufacturing sector and exports the most goods worldwide. It is also the second largest importer of products and has the fastest-growing consumer market in the world. Over the past several decades, China's economy has exhibited astounding development, propelling it to the position of second biggest economy in the world. When China began its economic reform program in 1978, it had a nominal GDP of USD214 billion, placing it ninth in the world; 35 years later, it had increased to USD9.2 trillion, placing it second (Statista, 2022).

The Chinese economy is greatly impacted by the ESG agenda due to the high level of environmental impact of manufacturing plants and the increasing population, along with rising stakeholder awareness. This is further compounded by the country's integration into the global market. The People's Republic of China's 5-year development plans address environmental and climate change challenges. Chinese President Xi Jinping declared his ambition to achieve carbon neutrality by the year 2060 in a speech during the UN General Assembly's 75th session in 2020. The Chinese government started allocating a sizable amount of money during the 12th 5-year plan's time period (2011–2015), and the tendency persisted during the 13th 5-year plan. China set long-term climate targets in the 14th 5-year plan (2021–2025), including lowering CO_2 emissions (Asian Development Bank, 2021). By 2060, China plans to achieve carbon neutrality.

The corporate world in China discloses information about sustainability strategies, programs, and projects on a voluntary basis, except for several categories of companies. Firstly, the disclosure of environmental and social impacts and activities to minimize these impacts is mandatory for "key pollutants." Since 2022, companies trading on the Star Market of Shanghai Stock Exchange are required to disclose ESG information. All public companies will be required to disclose information about the impact on the environment after the creation of a single standard by 2025 by the Ministry of Ecology and Environment of the People's Republic of China (Ziying, 2022).

China's ESG investment market, while small, has developed rapidly in recent years. After the USA, the country presently has the second largest green bond market in the world. There were 1643 green bonds in China as of December 31, 2021, with a total market value of RMB 1727 billion (about US$270 billion) (Wu, 2022).

The analysis over Chinese companies shows the focus of the leading companies on the environmental issue, while government regulations and recommendations are the main drivers for sustainability. Sustainable corporate governance is highly developed with strategic approach to sustainability, environmental risk management, stakeholder communication, and a set of basic policies and codes in the ESG fields. Companies are transparent and present information on corruption and unethical cases and have anti-corruption measures and initiatives, as well as anti-corruption education programs for employees.

The following description is based on the research of ten companies registered in China that are presented in at least one ESG rating (Sustainalytics/MSCI/Dow Jones Sustainability Index) and have nonfinancial report for 2020 or 2021. Companies present each of the industries (according to GICS): energy, utilities, materials, industrials, financials, communication services, information technology, healthcare, consumer discretionary or consumer staples, and real estate. All companies are large (according to EU criteria). Most of the sample companies are non-governmental, listed on more than one exchange, where the most common exchange is the Hong Kong Stock Exchange.

These leading corporations state in their reports that one of the main reasons for actively following the ESG agenda, specifically the environmental agenda, is the national-level requirements set by the state to improve the economic situation and promote the development of green finance.

China's sustainable development agenda is a critical aspect of the country's socioeconomic development. In the current 5-year plan (2021–2025), priorities include climate change, environmental protection, and reducing inequality between urban and rural areas. The objective of achieving carbon neutrality by 2060 is driving the demand for better management of ESG risks and increased disclosure of information. Companies are required to disclose greenhouse gas emissions, especially those from "key pollutants" list.

Nine out of ten companies have an ESG strategy. The strategy usually gives priority to the environmental direction of activity, and only three companies have goals in all three directions of E, S, and G. For example, China International Marine Containers (Group) Co., Ltd. outlines four directions of sustainability aims in its sustainable development strategy: operational excellence, digital transformation, human resources, risk management, capital, and resources. Out of the aims, the social aim is presented as "optimize the system and mechanism, and enrich and improve 5S control to create common cause," and one of the environmental aims sounds like "to give great impetus to the green and low-carbon development and gradually move toward the goal of emission peak and carbon neutrality." The company has outlined two to four measures for each of these sustainability aims. For instance, under the social aim, the company aims to achieve a 20% decrease in the number of working days lost due to work-related injuries by 2023, as compared to the baseline year of 2016. By 2025, the target is to achieve a 30% decrease in such days lost. As of 2019, the company had already achieved a 10% reduction in lost working days per year (CIMC, 2022).

Companies take into consideration ESG risks, including the possible changes in environmental and social legislation, threats to their reputation due to negative environmental impacts, risks associated with unstable markets, and changes in raw material prices. The risks are usually described in a balanced manner, but more importance is given to environmental and climate risks.

All companies have a set of top-level documents regarding ESG aspects, for example, code of conduct, carbon emission policy, or environmental policy.

The structure of the board for the majority of the companies may be described as non-balanced and non-diversified. Independent directors are present in only half of the companies in the sample; one company has more than 50% independent directors. The presence of independent directors in the sample companies does not correlate with the form of ownership of the company. The percentage of female directors varies from 0% to 14%, which is considered a low percentage in comparison to balanced 50–50 expectations.

Companies take stakeholder requests onboard, and half of the companies integrate stakeholders into decision-making processes. Half of the companies have stakeholder or ESG committee that is responsible for building and overseeing ESG strategy, which includes managers from different function, or even invites outside professionals for consultations. The committee may be called Sustainable Development Committee, ESG committee, or Corporate Governance Committee that includes functions in the ESG field.

Example 8.2 ESG committee at Sands China

Sands China Ltd. is a Chinese company, a subsidiary of Las Vegas Sands Corporation (NYSE: LVS), that develops, owns, and operates integrated resorts and casinos in Macau.

Company has ESG committee in its corporate governance system. The ESG committee is mainly composed of independent nonexecutive directors and is responsible for the following:

- Monitoring and overseeing the group's ESG strategy and approach, the corporate ESG reporting, and the implementation of the ESG-related policies and initiatives.
- Reviewing the company's environmental and social performance.
- Reviewing and making recommendation to the board with regard to the ESG-related matters, including the annual ESG report.
- Evaluating its own performance and reviewing the adequacy of its terms of reference annually (Sands China Ltd., 2021).

Example 8.3 "PetroChina Open Day" for stakeholders
PetroChina Company Limited is the listed arm of the state-owned China National Petroleum Corporation, based in Beijing's Dongcheng District. The firm is currently Asia's top producer of oil and gas.

Being a large corporation, PetroChina has a wide list of stakeholders. Apart from day-to-day communication and transparency, the company also organizes open days where representatives of stakeholder groups can visit the company in person and see its operations from the inside. In 2021, PetroChina invited nearly a hundred affiliated entities to the 4th "PetroChina Open Day" and invited approximately 10,000 officials from local governments, media reporters, customer representatives, residents from surrounding communities, and students to visit the company and get a firsthand look at it. PetroChina provided the public with on-site experiences, communication, and other activities to showcase its scientific and technological advances, green low-carbon achievements, and social responsibility efforts (PetroChina Ltd., 2022).

With regard to anti-corruption mechanisms, over half of the companies have implemented a code of ethics, an anti-corruption policy, and anti-corruption measures and provide training on anti-corruption to employees. In addition, a hotline for reporting corruption is in operation. Furthermore, 40% of the companies have an ethics committee. Among the ten companies, eight of them disclose the number of corruption cases that have occurred during the reporting period.

Example 8.4 Third-party hotline for anti-corruption
Continental Gold, a subsidiary of Zijin Mining Group Co., Ltd., has launched a third-party infringement-reporting platform to protect the legitimate rights and interests of whistleblowers.

The program aims to ensure the secure and confidential transmission of reports of violations. The information is processed by a third-party platform, significantly increasing its confidentiality and protecting whistleblowers from negative consequences. This provides valuable information for Continental Gold to combat internal and external collusion, illegal mining, theft of precious metals, and more. This not only protects the company but also safeguards the legitimate rights and interests of whistleblowers through a legally protected third-party platform (Zijin Mining Group, 2022).

8.2.2 *Indian Case*

India is ranked the seventh largest economy and third largest in terms of purchasing power parity (PPP). The actions taken by India to attain the SDGs are of great significance not only to India but also to the rest of the world, given that India has the second largest population in the world. The current sustainability strategy for India calls for greater attention to social and environmental issues like heavy dependence on coal, high poverty rates, significant undernourishment of the population, gender and income inequality, significant gaps in education and health, unbalanced urbanization, and an increase in the frequency of droughts and floods (Khalid et al., 2021).

India is voicing ambitious plans in the field of sustainable development: country has set a goal to achieve carbon neutrality by 2070, reduce the carbon intensity of the economy, and increase the share of renewable energy by 2030. India became one of the first countries where corporate social responsibility (CSR) was made mandatory. According to the Companies Act (2013) issued by the Ministry of Corporate Affairs, large companies are required to disclose information about CSR and direct funds to CSR in the amount of 2% of the average net profit for 3 financial years. Moreover, 1000 largest public companies are required to publish ESG reports in the format established by the regulator (ClearTax, 2022; World Economic Forum, 2022).

The following description is presenting a short case review of sustainable governance practices and motivation behind these practices for ten companies registered in India. The companies are leaders in ESG development: they are present at one or more ESG ratings or indexes (Sustainalytics/Morningstar/S&P/Dow Jones) and have sustainability/integrated or similar public reporting. Listing on the National Stock Exchange of India or Bombay Stock Exchange implies some obligations in terms of transparency and management rules for these companies. Companies are from all main industries (according to GICS): energy, utilities, materials, industrials, financials, communication services, information technology, healthcare, consumer discretionary or consumer staples, and real estate.

The motivation for inclusion in the ESG agenda is primarily driven by the state's desire to introduce basic practices and standards to facilitate further interaction between national and supranational institutions. Attracting financing also plays a crucial role.

The disclosure of the contribution toward sustainable development goals, analysis of terminology, and descriptions of ESG directions show that India places the most attention on environmental (including climate) and social aspects of ESG. Water protection is one of the most popular material topics, possibly due to seasonal drying up of rivers in India and the increasing need to address the issue of climate change globally. Specific material issues addressed in the reports include cybersecurity and innovation, overpopulation and rehabilitation, as well as sustainable procurement.

Example 8.5 Information security at Coal India

Coal India Limited (CIL) is an Indian central public sector organization owned by the Government of India's Ministry of Coal. It is the world's largest government-owned coal producer. With over 290,000 people, the company plays a pivotal role in the socio-economic canvas of India.

(continued)

Example 8.5 (continued)

CIL puts large emphasis on IT technologies. According to the company, information technology is one of the most important business accelerators, providing new benefits to corporate operations. Since company is large in terms of employees, IT technologies are mostly understood as the tools for implementing HR aims. CIL gives its staff IT resources to help them be more efficient and productive. These resources are intended to be the tools for accessing and processing information relevant to particular areas of expertise. These materials assist officials in staying informed and doing their duties efficiently and effectively.

For keeping an IT system productive and safe, CIL has a cyber-security policy that attempts to safeguard information and information infrastructure from cyberattacks by utilizing a combination of processes, rules, technology, and collaboration. CIL has business contingency plans and incident response protocols in place, which are tested on a semiannual basis.

Apart from standard governance mechanisms applied to IT (policies, norms, plans), CIL has established Data Centre and Disaster Recovery Centre for ERP implementation. Security system of data center is regularly audited, and it is equipped with state-security devices including DDoS, WAF, and next generation firewall (Coal India, 2022).

ESG strategic approach descriptions are balanced for half of the companies, where E, S, and G aims and objectives are described with the same amount of detail and data. However, for some companies, it is not clear whether the strategic approach is supported by the board's accepted strategy or just a verbal expression of a desirable ESG approach. The most attention is usually paid to the environmental direction when describing goals. Among the stated environmental goals, the most common are increasing energy efficiency, achieving zero emissions, creating "green" products, and preserving natural resources. Social goals are mostly formulated as overcoming gender inequality, employee engagement, diversity and inclusiveness, and health and safety. In terms of corporate governance goals, companies most often voice compliance with internationally accepted management standards (e.g., ISO) and the use of innovation in production.

Regulation of ESG activities of the ten companies under consideration is carried out through a number of regulatory documents, including policies on sustainable development in a broad sense, as well as narrowly focused policies such as ecology, climate change, anti-corruption, human rights, health and safety, and codes of conduct for employees.

All of the reviewed companies have a risk management committee. Each company identifies risks in three areas: environmental, social, and management. Environmental risks most often indicated by companies are climate change, regulation of emissions, access to water and resources, and environmental disasters. Social risks are commonly described in terms of pandemic risks and difficulties in developing and retaining talent. In the field of management, companies see the risks of increased competition in the market, dependence on one type of product, and the emergence of new regulatory requirements.

The board of directors of Indian companies is more diverse than that of Chinese companies due to a higher percentage of independent directors, but the gender composition is unbalanced, similar to the Chinese case. Most companies' boards of directors are represented by independent directors, with the percentage being at least 50%. However, women make up less than 50% of the board, ranging from 0 to 42%.

Board committees responsible for providing recommendations on the implementation of sustainable development principles in the company's activities are present in nine out of ten companies. Four out of ten companies have a specialized ESG committee (or similar) that reports to the board. The main focus of these committees is to conduct independent assessments of companies' operations to ensure compliance with ESG principles. They are also responsible for developing and overseeing ESG strategies and communicating with external stakeholders on ESG topics.

Example 8.6 ESG committee at IndusInd Bank Limited

IndusInd Bank Limited is an Indian financial service headquartered in Mumbai, with total assets of approximately US$50 billion. Bank offers a list of products and services for individuals and corporate and government entities and states that sustainability is embedded into the core of the banking system. The peculiar characteristic of sustainability activities of a bank is that it has multiplier effect since it spreads its practice to counterparties of a bank, such as lenders, corporate and private investors, and others.

Bank has CSR and Sustainability Committee at the board level that reviews and evaluates the sustainability strategy of the Bank and provides an industry perspective on the sustainability agenda of the Bank, including climate change. This includes evaluating the Bank's response to climate change through its operations and assessing the Bank's performance against its ESG targets. The CSR and Sustainability Committee is comprised of members of the board including the managing director and CEO of the Bank.

Going down in the corporate hierarchy, the Sustainability Council is operating under the supervision of the CSR and Sustainability Committee. Sustainability Council is responsible for the Bank's sustainability strategy, goals, and performance. It helps align the Bank's sustainability policies with the operations of the Bank's business units. The Council also reviews the Bank's compliance with respect to sustainability and climate change.

The Bank has also recently established the Environment and Social Management System (ESMS) Committee, which expands corporate norms to the bank's counterparties. The ESMS Committee reviews and monitors the implementation of the ESMS function within the Bank, which is responsible for

(continued)

Example 8.6 (continued)
assessing environmental and social risks arising from the Bank's investments. The system includes an exclusion list that protects the Bank from investing in specific industries that have an extremely high social or environmental impact. The risk profile considered through the policy includes risks associated with global warming and climate change (IndusInd Bank, 2022).

The main anti-corruption mechanisms in Indian companies are the application of the Code of Ethics, Anti-corruption Policy, hotline, and professional educational meetings for employees.

Acknowledgments Non-commercial partnership Agency "Da-Strategy" provided data on sustainable corporate governance of companies from emerging markets.

References

Ausan, A. (2021, December 3). Russian Institute of Directors. RID. http://rid.ru/materials/izdatelstvo-delibri-gruppa-kompanij-ripol-klassik-gotovit-k-vyxodu-novuyu-knigu-igorya-belikova-sovet-direktorov-kompanii-novyj-podxod

Asian Development Bank. (2021). *The 14th Five-Year Plan of the People's Republic of China—Fostering High-Quality Development*. Retrieved May 07, 2023, from https://doi.org/10.22617/BRF210192-2

Cardoni, A., Kiseleva, E., & Lombardi, R. (2020). A sustainable governance model to prevent corporate corruption: Integrating anticorruption practices, corporate strategy and business processes. *Business Strategy and the Environment, 29*(3), 1173–1185. https://doi.org/10.1002/bse.2424

CIMC. (2022). *Corporate Social Responsibility & Environmental, Social and Governance Report 2021*. Retrieved May 06, 2023, from https://www.cimc.com/en/index.php?m=content&c=index&a=lists&catid=69

ClearTax. (2022, October 20). *Corporate social responsibility under section 135 of Companies Act 2013*. Retrieved May 06, 2023, from https://cleartax.in/s/corporate-social-responsibility

Coal India. (2022). *Business Responsibility and Sustainability Report—FY 2021–22*. Retrieved May 06, 2023, from https://www.coalindia.in/media/documents/CIL_-_BRSR_2021-22.pdf

Cornell, B., & Damodaran, A. (2020). Valuing ESG: Doing good or sounding good? *NYU Stern School of Business*. SSRN Journal. https://doi.org/10.2139/ssrn.3557432

Corruption Perceptions Index 2018. (2019). www.Transparency.Org. Retrieved 27 March 2019, from https://www.transparency.org/cpi2018

Crotty, J. (2016). Corporate social responsibility in The Russian Federation: A contextualized approach. *Business & Society, 55*(6), 825–853. https://doi.org/10.1177/0007650314561965

Donaldson, T., & Dunfee, T. W. (1994). Toward a unified conception of business ethics: Integrative social contracts theory. *Academy of Management Review, 19*(2), 252–284. https://doi.org/10.5465/amr.1994.9410210749

Donaldson, T., & Dunfee, T. W. (2002). Ties that bind in business ethics: Social contracts and why they matter. *Journal of Banking & Finance, 26*(9), 1853–1865. https://doi.org/10.1016/S0378-4266(02)00195-4

Garanina, T., & Muravyev, A. (2020). The gender composition of corporate boards and firm performance: Evidence from Russia. *Emerging Markets Review, 48*, 100772. https://doi.org/10.1016/j.ememar.2020.100772

Gazprom. (2022). Retrieved May 06, 2023, from https://www.gazprom.com/about/

Ziying, S. (2022). How should businesses in China react to new environmental disclosure requirements? *China Dialogue*. Retrieved May 06, 2023, from https://chinadialogue.net/en/business/how-should-businesses-in-china-react-to-new-environmental-disclosure-requirements/

IndusInd Bank. (2022). *Integrated-Report-2020-21*. Retrieved May 07, 2023, from https://www.indusind.com/content/dam/indusind-corporate/generic/IndusInd-Bank-Integrated-Report-2020-21.pdf

Ioannou, I., & Serafeim, G. (2019). *Corporate sustainability: A strategy?* Harvard Business School. SSRN Journal. https://doi.org/10.2139/ssrn.3312191

Khalid, A. M., Sharma, S., & Dubey, A. K. (2021). Concerns of developing countries and the sustainable development goals: Case for India. *International Journal of Sustainable Development & World Ecology, 28*(4), 303–315. https://doi.org/10.1080/13504509.2020.1795744

Kuznetsov, A., & Kuznetsova, O. (2003). Institutions, business and the state in Russia. *Europe-Asia Studies, 55*(6), 907–922. https://doi.org/10.1080/0966813032000123079

Mccarthy, D. J., & Puffer, S. M. (2008). Interpreting the ethicality of corporate governance decisions in Russia: Utilizing integrative social contracts theory to evaluate the relevance of agency theory norms. *Academy of Management Review, 33*(1), 11–31. https://doi.org/10.5465/amr.2008.27745006

Orlitzky, M. (2015). The politics of corporate social responsibility or: Why Milton Friedman has been right all along. *Annals in Social Responsibility, 1*(1), 5–29. https://doi.org/10.1108/ASR-06-2015-0004

PetroChina Ltd. (2022). *2021 Environmental, Social And Governance Report*. Retrieved May 07, 2023, from http://www.petrochina.com.cn/ptr/xhtml/images/2021esgen.pdf

Porter, M. E., & Kramer, M. R. (2006). Strategy and society: The link between competitive advantage and corporate social responsibility. *Harvard Business Review, 23*(5), sd.2007.05623ead.006. https://doi.org/10.1108/sd.2007.05623ead.006

Sands China Ltd. (2021). *2020 Environmental, Social & Governance Report*. Retrieved May 07, 2023, from https://assets.sandsresortsmacao.cn/content/sandschina/community-affairs/ESG-report/2020-Sands-China-ESG-report_EN.pdf

Statista. (2022). *China: GDP at current prices 1985–2027*. Retrieved May 06, 2023, from https://www.statista.com/statistics/263770/gross-domestic-product-gdp-of-china/

Ting, I. W. K., Azizan, N. A., Bhaskaran, R. K., & Sukumaran, S. K. (2020). Corporate social performance and firm performance: Comparative study among developed and emerging market firms. *Sustainability, 12*(1), 1. https://doi.org/10.3390/su12010026

Tumanov, S. V. (2021). *Anti-corruption policy of the Gazprom EP international B.V. Group Companies*. Retrieved May 06, 2023, from https://www.gazprom-international.com/d/textpage/74/116/anti-corruption-policy.pdf

World Economic Forum. (2022). *These are the challenges facing India's net-zero target*. Retrieved May 07, 2023, from https://www.weforum.org/agenda/2022/09/net-zero-challenges-india-target/

Wu, Y. (2022). China's green finance market: Policy support & investment opportunities. *China Briefing News*. Retrieved May 06, 2023, from https://www.china-briefing.com/news/chinas-green-finance-market-policies-incentives-investment-opportunities/

Zhao, M. (2012). CSR-based political legitimacy strategy: Managing the state by doing good in China and Russia. *Journal of Business Ethics, 111*(4), 439–460. https://doi.org/10.1007/s10551-012-1209-6

Zijin Mining Group. (2022). *2021 Environmental, Social and Governance Report*. Retrieved May 06, 2023, from https://www.zjky.cn/upload/file/2022/06/09/9b7954d122554e8ab7327852dc7d4004.pdf

Chapter 9
The Dependencies between Sustainable Governance and Market Value

9.1 Is Sustainable Governance Financially Sustainable?

Going from the theoretical definitions and case studies in previous chapters, governance is supposed to play a key role in comprehending the interactions between financial markets and firms' strategies and especially corporate sustainability.

Agency, institutional, and stakeholder theories suggest that sustainable governance has an impact on financial results. Companies have moved from debating whether to conduct corporate social responsibility or not to how to properly manage it as an ally of financial performance (Lu et al., 2018). Nonetheless, these theories describe the characteristic of effective corporate governance and the mechanisms of its influence on company's financial performance in a distinct way.

According to basic agency theory (Hölmstrom, 1979; Jensen & Meckling, 1976; Ross, 1973), the relationship among shareholders and the board of directors is considered an agency relationship, in which shareholders delegate the realization of corporate activities to the board of directors. The actions or reactions of the board of directors underline the information asymmetry between firm managers and financial markets (Eisenhardt, 1989). The agency theory of the firm suggests that when individuals engage in firm relationships, they are utility maximizers, self-seeking, and opportunistic, and, therefore, the governance system introduces mechanisms that align the interests of principals (owners) with those of their agents (the mangers) (Aguilera, 2005; Jensen & Meckling, 1976).

The main argument for examining the relative importance of governance in market value according to agency theory is that monitoring mechanisms are able to reduce agency problems between owners and managers and to reduce costs related to this problem (Caixe & Krauter, 2014; Fu, 2019; Worokinasih, 2020). In this sense, the vast literature proposes the evidence of a relationship between better governance practices and higher market value (Ammann et al., 2011; Black et al., 2006; Caixe & Krauter, 2014; Durnev & Kim, 2005; Worokinasih, 2020).

A. Cardoni, E. Kiseleva, *Sustainable Governance*, CSR, Sustainability, Ethics & Governance, https://doi.org/10.1007/978-3-031-37492-0_9

Institutional theory (DiMaggio & Powell, 1983; Meyer & Rowan, 1977; Tolbert & Zucker, 1999) is supposed to be more relevant for the modern economic systems (Aguilera, 2005). The institutional theory defines a set of formal and informal rules that affect business activity (North, 2010). According to them, the corporate organization is a social system or collective entity with pluralist interests and some common goals (Aguilera, 2005). A firm's right to exist is legitimized if its value system is consistent with that of the larger social system of which it is a part but is threatened when there is actual or potential conflict between the two value systems (Suchman, 1995).

According to this paradigm, corporate governance helps in incorporating external norms and rules into operations and structures, which enables the legitimacy and social acceptance (Briano-Turrent & Rodríguez-Ariza, 2016; DiMaggio & Powell, 1983). Therefore, corporate governance can be considered as a crucial factor for economic growth and financial stability (Aguilera, 2005).

The stakeholder theory (Donaldson & Preston, 1995; Freeman, 1984; Mitchell et al., 1997) with its interpretation of corporate governance is gaining the largest attention nowadays (Aguilera, 2005). According to it, corporate governance is responsible not only for protecting and increasing shareholder wealth but also for ensuring that strategic decisions are beneficial for all other stakeholders (Allais et al., 2017; Shahzad et al., 2016).

Such stakeholder-oriented or sustainable corporate governance focuses on the value-creating relationships with all stakeholders, avoiding negative impacts on the environment and society (Lombardi et al., 2019; Sánchez et al., 2011). Furthermore, it is expected that the expansion of corporate governance to include the firm's key stakeholders should not only be guided by regulatory or legal requirements but also encompass an ethical dimension (Lombardi et al., 2019). By this, stakeholder orientation has the potential to create a competitive advantage for proactive companies both from the possibility of being rewarded by the market and from avoiding risks (Allais et al., 2017).

9.2 Empirical Arguments for Positive Monetary Influence of Sustainable Corporate Governance

For verifying whether sustainable corporate governance has significant and positive impact on financial results, the authors undertook the analysis of 110 public companies with leading ESG results (Cardoni et al., 2022).

The sample was selected from the GRI Database, out of publicly listed companies from seven countries: the USA, the UK, Italy, Russia, Brazil, Australia, and South Africa. The countries represent all the continents and cover the various governance law enforcements (Anglo-Saxon and others), as well as level of development (developed countries, developing, and in transition). The companies are also

Table 9.1 Companies under analysis

Country	Number of companies	Industry	Number of companies
USA	71	Basic materials	17
Italy	12	Financial services	14
UK	10	Industrials	13
Russian Federation	9	Energy	12
Brazil	5	Real estate	12
Australia	2	Utilities	10
South Africa	1	Consumer cyclical (packaging and containers, apparel, gambling)	9
Anglo-Saxon	83	Healthcare	7
Other	27	Consumer defensive (farm products, packaged food, household and personal products)	6
Developed	95	Communication services	5
In transition	9	Technology	5
Developing	6	Sensitive to corruption[a]	57
		Not sensitive to corruption	53

Source: Author constructed
[a]According to Kottasova (2014)

Table 9.2 Dependent variable

Variable	Type	Description	Literature	Data source
Ymbv	Continuous	Market to book value, proportion	Siddiqui (2015)	Thomson Reuters data

Source: Author constructed

from different sectors and industries (Table 9.1), which enabled the authors to obtain generalizable results that are applicable in any context.

In order to analyze the level of sustainable governance, we used sustainability and annual reports published in English during 2018–2020 (the reporting period being 2018). Reports (with proxy statements as their supplements) were used for analysis as these are the most popular tools among companies for sharing sustainability (economic, social, and environmental) risk information and performance with stakeholders (Du et al., 2017; Dumay & Hossain, 2019). Thomson Reuters data were used as a source of financial data.

9.2.1 Dependent Variable

MBV (Table 9.2) was used as a reflection of the long-term value of the firm and market perceptions about the value of corporate governance.

9.2.2 Independent Variables

Independent variables were constructed based on the methodology described in the following book and largely influenced by the WEF 2020 Consultation Draft prepared by Deloitte, EY, KPMG, and PwC (World Economic Forum, 2020).

We constructed nine indicators as determinants of sustainable governance practice, where each indicator takes on a value from 0 to 1. The indicators were calculated for each company on the basis of sustainability and annual reports (together with proxy statements). Codification was undertaken manually by analyzing the relevant parts of the reports. The selective cross-check was undertaken for ensuring the reliability of the scores.

Next, we matched the indicators to the three components of sustainable governance: ESG strategy and risk management, board composition and stakeholder engagement, and anti-corruption and business ethics. Transparency issue was integrated into each of the sections. Each component consists of three relevant indicators and takes on a value from zero to 3. Finally, a sustainable governance score was built as the sum of the three microsections. It goes from 0 to 9, where 0 is not sustainable governance and 9 is best sustainable governance.

Consequently, the independent variable was created (Tables 9.3 and 9.4).

Table 9.3 Independent variables

Variables	Type	Description	Literature	Source
SG	Continuous	Total sustainable governance score	Aguilera (2005), Bassen and Kovács (2008), Briano-Turrent and Rodríguez-Ariza (2016), Cardoni, Kiseleva and Lombardi (2020), Cardoni, Kiseleva and Taticchi (2020), Cheung et al. (2007), Ioannou and Serafeim (2019), Lombardi et al. (2019), Luo (2005), Porter and Kramer (2006), Sánchez et al. (2011), Simpson and Samson (2010), World Economic Forum (2020), Yadav and Pathak (2016)	Compiled by the authors based on GRI reports list and companies' annual and sustainability reports, as well as proxy statements

Source: Author constructed, based on GRI Reports List

Table 9.4 Core metrics and disclosures of the total sustainable governance score

Core metrics and disclosures	Calculation
ESG strategy and risk management component	
Setting purpose	Dichotomous variable that takes the value 1 if the company has a stated purpose linked to E, S, and G. 0—Otherwise or not disclosed
Integrating ESG risks and opportunities into business process	Variable that takes the value of 0–1, where each of the following points is assigned a value of 0.5: - Risks from E, S, and G spheres. - Inclusion of emerging principal risks of fourth industrial revolution (data security/digitalization, etc.)
Disclosure about integration of ESG risks and opportunities into business process	Variable that takes the value of 0–1, where each of the following points is assigned a value of 0.25: - Disclosure of company-level ESG risk factors as opposed to generic sector risks. - Disclosure of the board appetite in respect of these risks. - Disclosure of how these risks have moved over-time. - Disclosure of the response to risks.
Board composition and stakeholder engagement component	
Board composition	Variable that takes the value of 0–1, where each of the following points is assigned a value of 0.25 No COB-CEO duality: - Independence of the majority of the board. - Gender diversity and/or membership of under-represented social groups. - Stakeholder representation in the board or committees.
Board composition disclosure	Variable that takes the value of 0–1, where each of the following points is assigned a value of 0.(33): - Disclosure of tenure on the governance body. - Disclosure of the number of each individual's other significant positions and commitments and the nature of the commitments. - Disclosure of competencies relating to economic, environmental, and social topics (at least one).
Impact of material issues on stakeholders	Variable that takes the value of 0–1, where each of the following points is assigned a value of 0.5: - The material topics are identified. - Stakeholders were questioned during this process.
Anti-corruption and business ethics component	
Anti-corruption	Variable that takes the value of 0–1, where each of the following points is assigned a value of 0.(33): - Governance body members have received training on the organization's anti-corruption policies and procedures. - Employees have received training on the

(continued)

Table 9.4 (continued)

Core metrics and disclosures	Calculation
	organization's anti-corruption policies and procedures. - Business partners have received training on the organization's anti-corruption policies and procedures.
Anti-corruption disclosure	Variable that takes the value of 0–1, where each of the following points is assigned a value of 0.(33): - Disclosure of total number and nature of incidents of corruption confirmed during the current year, but related to previous years. - Disclosure of total number and nature of incidents of corruption confirmed during the current year, related to this year. - Disclosure of the number of employees/governance members or business partners.
Protected ethics advice and reporting mechanisms	Variable that takes the value of 0–1, where each of the following points is assigned a value of 0.5: - Company has internal and external mechanisms for seeking advice about ethical behavior (politics, norms, standards, events, etc.). - Company has internal and external mechanisms for reporting concerns about unethical behavior (hotline and the like).

Source: Own construction

9.2.3 Control Variables

The control variables list was constructed based on the MBV formula (Cheung et al., 2007) and the modified Ohlson (1995) model of market value (de Klerk & de Villiers, 2012; Hassel et al., 2005). The most relevant variables for our research were selected based on the literature on closely related topics (Table 9.5). In addition to company-level factors, the institutional framework also affects corporate governance compliance (Aguilera & Jackson, 2010). Therefore, we included controls for country and industry.

9.2.4 Results

We found out that Ymbv and SG may have a nonlinear correlation; thus, we created two variables for two intervals of SG (Fig. 9.1).

The borderlines for the intervals were determined based on the peaks of locally weighted regressions for *Ymbv-SG*. Variable *SG_begin* is supposed to represent beginners with lowest sustainable governance scores, while variable *SG_advance* shows the best practice with the highest scores. Adopting the piecewise regression

Table 9.5 Control variables

Variables	Type	Description	Literature	Source
General variables				
ROA	Continuous	Pretax ROA, share	Cheung et al. (2007), de Klerk and de Villiers (2012), Hassel et al. (2005)	Thomson Reuters data
Div	Continuous	Dividend yield, share	de Klerk and de Villiers (2012), Hassel et al. (2005)	Thomson Reuters data
BVPS	Continuous	Book value per share, USD	de Klerk and de Villiers (2012), Hassel et al. (2005), Qureshi et al. (2020)	Thomson Reuters data
Size				
TA	Continuous	Natural logarithm of total assets	Cheung et al. (2007), de Klerk and de Villiers (2012), Hassel et al. (2005)	Thomson Reuters data
Industry				
Ind	Dummy	Industry sensitivity to corruption. Dummy variable that takes on 1 for sensitive industries (extraction, construction, transportation, information/communication, manufacturing), 0 otherwise	Cahan et al. (2016), Kottasova (2014), Qureshi et al. (2020), Xie et al. (2019)	Thomson Reuters data
Country				
Dev	Categorical	Developing/developed countries. Dummy variable that takes on 1 for developed countries, 0.5 for economies in transition, and 0 for developing countries	Aguilera and Jackson (2010), United Nations (2019), Xie et al. (2019)	World Bank data

Source: Own construction with data from Thomson Reuters and the World Bank

approach and differentiating between the *beginners* and the *advanced* companies, the theoretical hypothesis was tested using two statistical models (Table 9.6).

Model 1 shows a significant ($p < 0.05$) and positive (coefficient = 1.45) dependence between *Ymbv* and *SG_begin*. In other words, market valuation increases when companies with an initially low sustainable governance score (or *beginners* in sustainable governance) improve their sustainable governance practice. The model has the highest explanatory ability with an R^2 of 88%.

Model 2 explores the role of sustainable governance for the upper interval of sustainable governance scores (*SG_advance*). The explanatory ability of this model is poor in this interval ($R^2 = 0.41$), and only two control variables are significant (*ROA* and *TA*). Contrary to Model 1, sustainable governance is insignificant and has

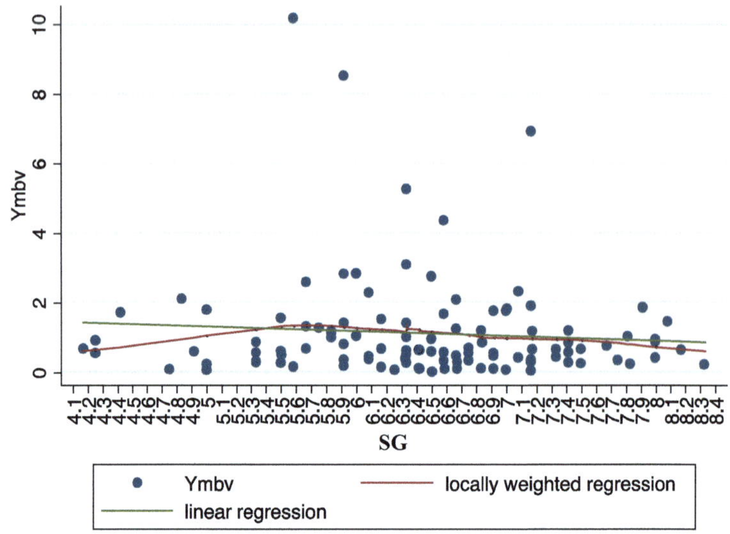

Fig. 9.1 Linear and locally weighted regressions for *Ymbv* and *SG*. Source: own construction

Table 9.6 Comparison of the models

	Model 1	Model 2
Sustainable governance variables		
SG_begin	1.429* (0.529)	.
SG_advance	.	−0.552 (0.286)
Control variables		
ROA	15.487*** (2.286)	5.900** (1.712)
Div	−45.246** (14.650)	−8.646 (9.060)
BVPS	−41.344** (16.953)	−0.128 (0.260)
TA	0.625 (0.297)	−0.210* (0.096)
Dev	3.603** (1.128)	0.600 (0.666)
Ind	−0.500 (0.727)	−0.635 (0.317)
Number of observations	21	61
p	0.0000	0.0001
R^2	0.879	0.4082
VIF	4.00	1.51
AIC	62.118	199.246
BIC	70.475	216.133

Standard errors are indicated in parentheses
Significance levels: $*p < 0.05$; $**p < 0.01$; $***p < 0.001$
Source: Authors' calculations

a negative sign (coefficient = −0.44). Thus, it appears that the market value does not increase with the further development of sustainable governance from an already high level (*SG_advance*). In fact, the graph suggests that the influence may be even negative.

A positive link was found for the sample of companies in the interval of low level of sustainable governance (Model 1). This interval represents the *beginners*: companies at basic levels of sustainable governance mechanisms. Conversely, the results of Model 2 suggest that investors do not react appreciably to developments in sustainable governance when it is already scored highly (i.e., for the sample of companies with developed sustainable governance and anti-corruption programs).

Changes in investors' attitude are therefore implied to vary, depending on the levels of maturity of the sustainable governance. For instance, investors add a company to their portfolio and pay more for its stock when the company expresses an intention to improve its poor sustainable governance practice. However, when there is clarity on a company's sustainable governance development path, investors pay no further attention to any changes in the sustainable governance scores, as these changes have already been anticipated. These findings are consistent with the results obtained by several authors, where unexpected CSR disclosure increased market value, while expected components had no influence (Cahan et al., 2016). Qureshi et al. (2020) contend that the corporate governance systems that operate in an already developed governance arena may not provide any additional incentive for the market to have an effect on the price. Cornell and Damodaran suggest the same dependence, explaining it by reference to market equilibrium. According to them, there is an adjustment period during which highly rated ESG stocks will outperform low ESG stocks; however, once an equilibrium point is reached, the expected returns offered by the former stocks decline (Cornell & Damodaran, 2020).

Based on stakeholder theory, the weak results of *ESG-MBV* correlation could be justified by the fact that bad corporate sustainability news are anticipated by stakeholders more than good news (Flammer, 2012). Indeed, one corruption case leads to more detrimental and visible effects than one more corruption-free year. Hence, sustainable governance actions to avoid negative events (i.e., those that eliminate corrupt transactions and reduce preliminary ethical risks) are appreciated more than advances in the area.

From the perspective of institutional theory, the reasons of poor results for developed companies may lie in stricter ethical corporate governance norms and regulations (Agatiello, 2008; Cardoni, Kiseleva, & Lombardi, 2020), and with each new requirement, companies bounce back to beginner status. Thus, companies struggle with primary compliance to new laws, while best practice becomes too expensive and difficult to achieve in such a stringent environment.

Moreover, the standard neoclassical assumption of decreasing marginal returns to production factors may explain the results in the most comprehensible manner (Flammer, 2012; Hollander, 1997). According to this premise, in the early stages of sustainable governance development, it is relatively simple and inexpensive to improve the level. As a company's ethical governance system advances, it may become progressively more challenging and costly to make further progress, with changes resulting in the same or even fewer benefits. Our research indicates that the market rewards a basic level of sustainable governance policies. In fact, investors value the risk mitigation resulting from the implementation of sustainable governance mechanisms over opportunities for value creation.

References

Agatiello, O. R. (2008). Ethical governance: Beyond good practices and standards. *Management Decision, 46*(8), 1132–1145. https://doi.org/10.1108/00251740810901345

Aguilera, R. V. (2005). Corporate governance and director accountability: An institutional comparative perspective. *British Journal of Management, 16*(s1), 39–53. https://doi.org/10.1111/j.1467-8551.2005.00446.x

Aguilera, R. V., & Jackson, G. (2010). Comparative and international corporate governance. *Academy of Management Annals, 4*(1), 485–556. https://doi.org/10.1080/19416520.2010.495525

Allais, R., Roucoules, L., & Reyes, T. (2017). Governance maturity grid: A transition method for integrating sustainability into companies. *Journal of Cleaner Production, 140*, 213–226. https://doi.org/10.1016/j.jclepro.2016.02.069

Ammann, M., Oesch, D., & Schmid, M. M. (2011). Corporate governance and firm value: International evidence. *Journal of Empirical Finance, 18*(1), 36–55. https://doi.org/10.1016/j.jempfin.2010.10.003

Bassen, A., & Kovács, A. M. (2008). Environmental, social and governance key performance indicators from a capital market perspective. *Zeitschrift für Wirtschafts- und Unternehmensethik, 9*(2), 182–192.

Black, B. S., Jang, H., & Kim, W. (2006). Does corporate governance predict firms' market values? Evidence from Korea. *The Journal of Law, Economics, and Organization, 22*(2), 366–413. https://doi.org/10.1093/jleo/ewj018

Briano-Turrent, G. d. C., & Rodríguez-Ariza, L. (2016). Corporate governance ratings on listed companies: An institutional perspective in Latin America. *European Journal of Management and Business Economics, 25*(2), 63–75. https://doi.org/10.1016/j.redeen.2016.01.001

Cahan, S. F., De Villiers, C., Jeter, D. C., Naiker, V., & Van Staden, C. J. (2016). Are CSR disclosures value relevant? Cross-country evidence. *European Accounting Review, 25*(3), 579–611. https://doi.org/10.1080/09638180.2015.1064009

Caixe, D. F., & Krauter, E. (2014). The relation between corporate governance and market value: Mitigating endogeneity problems. *Brazilian Business Review, 11*(1), 80–110. https://doi.org/10.15728/bbr.2014.11.1.5

Cardoni, A., Kiseleva, E., Arduini, S., & Terzani, S. (2022). From sustainable value to shareholder value: The impact of sustainable governance and anti-corruption programs on market valuation. *Business Strategy and the Environment*. https://doi.org/10.1002/bse.3328

Cardoni, A., Kiseleva, E., & Lombardi, R. (2020). A sustainable governance model to prevent corporate corruption: Integrating anticorruption practices, corporate strategy and business processes. *Business Strategy and the Environment, 29*(3), 1173–1185. https://doi.org/10.1002/bse.2424

Cardoni, A., Kiseleva, E., & Taticchi, P. (2020). In search of sustainable value: A structured literature review. *Sustainability, 12*(2), Article 2. https://doi.org/10.3390/su12020615

Cheung, Y.-L., Thomas Connelly, J., Limpaphayom, P., & Zhou, L. (2007). Do Investors really value corporate governance? Evidence from the Hong Kong market. *Journal of International Financial Management & Accounting, 18*(2), 86–122. https://doi.org/10.1111/j.1467-646X.2007.01009.x

Cornell, B., & Damodaran, A. (2020). Valuing ESG: Doing good or sounding good? *NYU Stern School of Business, SSRN Journal.* https://doi.org/10.2139/ssrn.3557432

de Klerk, M., & de Villiers, C. (2012). The value relevance of corporate responsibility reporting: South African evidence. *Meditari Accountancy Research, 20*(1), 21–38. https://doi.org/10.1108/10222521211234200

DiMaggio, P. J., & Powell, W. W. (1983). The iron cage revisited: Institutional isomorphism and collective rationality in organizational fields. *American Sociological Review, 48*(2), 147–160. JSTOR. https://doi.org/10.2307/2095101

Donaldson, T., & Preston, L. E. (1995). The stakeholder theory of the corporation: Concepts, evidence, and implications. *The Academy of Management Review, 20*(1), 65–91. JSTOR. https://doi.org/10.2307/258887

Du, S., Yu, K., Bhattacharya, C. B., & Sen, S. (2017). The business case for sustainability reporting: Evidence from stock market reactions. *Journal of Public Policy & Marketing, 36*(2), 313–330. https://doi.org/10.1509/jppm.16.112

Dumay, J., & Hossain, M. A. (2019). Sustainability risk disclosure practices of listed companies in Australia. *Australian Accounting Review, 29*(2), 343–359. https://doi.org/10.1111/auar.12240

Durnev, A., & Kim, E. H. (2005). To steal or not to steal: Firm attributes, legal environment, and valuation. *The Journal of Finance, 60*(3), 1461–1493. https://doi.org/10.1111/j.1540-6261.2005.00767.x

Eisenhardt, K. M. (1989). Agency theory: An assessment and review. *Academy of Management Review, 14*(1), 57–74. https://doi.org/10.5465/amr.1989.4279003

Flammer, C. (2012). Corporate social responsibility and shareholder reaction: The environmental awareness of investors. *Academy of Management Journal, 56*(3), 758–781. https://doi.org/10.5465/amj.2011.0744

Freeman, R. (1984). *Strategic management: A stakeholder approach.* Cambridge University Press.

Fu, Y. (2019). The value of corporate governance: Evidence from the Chinese anti-corruption campaign. *The North American Journal of Economics and Finance, 47*, 461–476. https://doi.org/10.1016/j.najef.2018.06.001

Hassel, L., Nilsson, H., & Nyquist, S. (2005). The value relevance of environmental performance. *European Accounting Review, 14*(1), 41–61. https://doi.org/10.1080/0963818042000279722

Hollander, S. (1997). *The economics of Thomas Robert Malthus.* University of Toronto Press.

Hölmstrom, B. (1979). Moral Hazard and observability. *The Bell Journal of Economics, 10*(1), 74–91. JSTOR. https://doi.org/10.2307/3003320

Ioannou, I., & Serafeim, G. (2019). Corporate sustainability: A strategy? *Harward Business School, SSRN Journal.* https://doi.org/10.2139/ssrn.3312191.

Jensen, M. C., & Meckling, W. H. (1976). Theory of the firm: Managerial behavior, agency costs and ownership structure. *Journal of Financial Economics, 3*(4), 305–360. https://doi.org/10.1016/0304-405X(76)90026-X

Kottasova, I. (2014, December 2). *World's most corrupt industries.* CNNMoney. https://money.cnn.com/2014/12/02/news/bribery-foreign-corruption/index.html.

Lombardi, R., Trequattrini, R., Cuozzo, B., & Cano-Rubio, M. (2019). Corporate corruption prevention, sustainable governance and legislation: First exploratory evidence from the Italian scenario. *Journal of Cleaner Production, 217*, 666–675. https://doi.org/10.1016/j.jclepro.2019.01.214

Lu, W., Ye, M., Chau, K. W., & Flanagan, R. (2018). The paradoxical nexus between corporate social responsibility and sustainable financial performance: Evidence from the international construction business. *Corporate Social Responsibility and Environmental Management, 25*(5), 844–852. https://doi.org/10.1002/csr.1501

Luo, Y. (2005). An organizational perspective of corruption. *Management and Organization Review, 1*(01), 119–154. https://doi.org/10.1111/j.1740-8784.2004.00006.x

Meyer, J. W., & Rowan, B. (1977). Institutionalized organizations: Formal structure as myth and ceremony. *American Journal of Sociology, 83*(2), 340–363. JSTOR.

Mitchell, R. K., Agle, B. R., & Wood, D. J. (1997). Toward a theory of stakeholder identification and salience: Defining the principle of who and what really counts. *Academy of Management Review, 22*(4), 853–886. https://doi.org/10.5465/amr.1997.9711022105

North, D. C. (2010). *Understanding the process of economic change*. https://press.princeton.edu/books/paperback/9780691145952/understanding-the-process-of-economic-change.

Ohlson, J. A. (1995). Earnings, book values, and dividends in equity valuation*. *Contemporary Accounting Research, 11*(2), 661–687. https://doi.org/10.1111/j.1911-3846.1995.tb00461.x

Porter, M. E., & Kramer, M. R. (2006). Strategy and society: The link between competitive advantage and corporate social responsibility. *Harvard Business Review, 23*(5), sd.2007.05623ead.006. https://doi.org/10.1108/sd.2007.05623ead.006

Qureshi, M. A., Kirkerud, S., Theresa, K., & Ahsan, T. (2020). The impact of sustainability (environmental, social, and governance) disclosure and board diversity on firm value: The moderating role of industry sensitivity. *Business Strategy and the Environment, 29*(3), 1199–1214. https://doi.org/10.1002/bse.2427

Ross, S. A. (1973). The economic theory of agency: The Principal's problem. *The American Economic Review, 63*(2), 134–139. JSTOR.

Sánchez, J. L. F., Sotorrío, L. L., & Díez, E. B. (2011). The relationship between corporate governance and corporate social behavior: A structural equation model analysis. *Corporate Social Responsibility and Environmental Management, 18*(2), 91–101. https://doi.org/10.1002/csr.244

Shahzad, A. M., Rutherford, M. A., & Sharfman, M. P. (2016). Stakeholder-centric governance and corporate social performance: A cross-National Study. *Corporate Social Responsibility and Environmental Management, 23*(2), 100–112. https://doi.org/10.1002/csr.1368

Siddiqui, S. S. (2015). The association between corporate governance and firm performance – A meta-analysis. *International Journal of Accounting and Information Management, 23*(3), 218–237. https://doi.org/10.1108/IJAIM-04-2014-0023

Simpson, D., & Samson, D. (2010). Environmental strategy and low waste operations: Exploring complementarities. *Business Strategy and the Environment, 19*(2), 104–118. https://doi.org/10.1002/bse.626

Suchman, M. C. (1995). Managing legitimacy: Strategic and institutional approaches. *Academy of Management Review, 20*(3), 571–610. https://doi.org/10.5465/amr.1995.9508080331

Tolbert, P. S., & Zucker, L. G. (1999). The institutionalization of institutional theory. In *Studying organization: Theory & Method* (pp. 169–184). SAGE Publications Ltd. https://doi.org/10.4135/9781446218556

United Nations. (2019). *World economic situation and prospects*. Retrieved May 05, 2023, from https://www.un.org/development/desa/dpad/wp-content/uploads/sites/45/WESP2019_BOOK-ANNEX-en.pdf

World Economic Forum. (2020). *Toward common metrics and consistent reporting of sustainable value creation*. Retrieved May 05, 2023, from https://www.weforum.org/whitepapers/toward-common-metrics-and-consistent-reporting-of-sustainable-value-creation/

Worokinasih, S. (2020). The mediating role of Corporate Social Responsibility (CSR) Disclosure on Good Corporate Governance (GCG) and Firm value. 14, 10.

Xie, J., Nozawa, W., Yagi, M., Fujii, H., & Managi, S. (2019). Do environmental, social, and governance activities improve corporate financial performance? *Business Strategy and the Environment, 28*(2), 286–300. https://doi.org/10.1002/bse.2224

Yadav, R., & Pathak, G. S. (2016). Young consumers' intention towards buying green products in a developing nation: Extending the theory of planned behavior. *Journal of Cleaner Production, 135*, 732–739. https://doi.org/10.1016/j.jclepro.2016.06.120

Chapter 10
Sustainable Governance and Crises

10.1 Previous Experience

Governance system is an equilibrium response to its operating characteristics and its competitive environment, which was postulated back in 1985 by Demsetz and Lehn (1985) and is becoming even more relevant nowadays. Indeed, corporate governance helps company to navigate both indoor activities and outdoor challenges.

One of the main outer challenges is economic crises: they appear in a cyclical way and are not always foreseeable. Crises bring destructive consequences for both individual companies and entire economies. This is particularly true in the context of corporate governance, which has historically seen significant changes and developments in response to crisis situations. Below, we outline some of the major changes that have occurred in corporate governance practice following crises.

The Great Depression of 1929–1939 is widely considered to have started the active and widespread debates around the topic of corporate governance. The crisis began in the USA with the stock market crash of 1929, which triggered a chain reaction, resulting in the ruin of many companies, banks, and investors. The consequences of the Great Depression affected American consumption, banking, and the economy as a whole, and its effects spread worldwide.

Following the stock market crash, governments created an institutional system to control and oversee financial markets, and a set of legislative papers were issued to regulate the rights and obligations of financial market participants.

In the USA, for instance, the Securities Act of 1933 was the first significant federal legislation governing the sale of securities. It is widely considered a disclosure law that mandates security issuers to provide information that enables investors to make informed investment decisions, thereby prohibiting the fraudulent sale of registered securities. Following the regulation of securities issuance, the government expanded its control to the secondary market, where securities are traded. The 1934 Securities Exchange Act focused on ensuring that exchanges, brokers, and dealers act in the best interests of investors. Establishing regulatory bodies was also a

A. Cardoni, E. Kiseleva, *Sustainable Governance*, CSR, Sustainability, Ethics & Governance, https://doi.org/10.1007/978-3-031-37492-0_10

crucial step: the Securities Commission and the National Association of Securities Dealers were created.

In the same years, the idea of corporate governance with dispersed ownership and separation of ownership and control was further developed and refined (Berle & Means, 1932). Berle and Means in their widely cited book *The Modern Corporation and Private Property* described the changing landscape of a company and presented a "corporate system" that evolved from it. They warned about the concentration of economic power in the hands of managers of the increasingly rising corporations while at the same time highlighting the trend toward a quasi-public role of corporations with multiple ownerships that should be navigated through corporate governance.

Already in 1976, the term "corporate governance" first appeared in the Federal Register, the official journal of the federal government in the USA. Next, the precise corporate governance legislation started to appear with the regulations over the board structure, duties of corporate directors, audit and nominating committees, shareholders' rights, and, of course, transparency.

For example, in 1976, the SEC prompted the New York Stock Exchange (NYSE) to require each listed corporation to have an audit committee composed of all independent board directors. Such norms soon became widespread over the world developing the corporate governance resilience to shocks and manipulations.

The Cadbury Committee was established by the London Stock Exchange in 1991, and in 1992, the Cadbury Report was issued. This report presented the code of best practice in the field of corporate governance with a "comply or explain" rule for its use by corporations.

International initiatives also emerged during this period. In 1999, the OECD Principles of Corporate Governance gained global recognition as a benchmark for sound corporate governance. These principles were developed to aid policymakers in assessing and enhancing the legal, regulatory, and institutional framework for corporate governance to promote economic efficiency, long-term growth, and financial stability.

Despite the expectations that the practice will spread to all the countries smoothly, *emerging economies* met challenges in the adoption process. Some countries with closed economies lacked the institutional infrastructure and legislative framework necessary for a market economy, free enterprise, and globalized trade. As a result, corporate management had tremendous power without adequate oversight.

In many emerging economies, the management has the ability to transfer cash and other assets out of the company with outside investors. These transfers could be made to pay personal debts, shore up another company with different shareholders, or be deposited directly into a foreign bank account. Furthermore, the fact that management in most emerging markets also acts as the controlling shareholder makes these transfers even easier to achieve. As a result, corporate downturns in these countries are often associated with a significant increase in the expropriation of cash and tangible assets by managers.

With such conditions on start, transition period from local economic regimes to an open market with fast-growing corporations has led to another one widely known crisis—the so-called Asian financial crisis of 1997–1998. The Asian financial crisis started as a currency crisis triggered by Bangkok's decision to unpeg the Thai baht from the US dollar. This led to a series of currency devaluations and capital outflows. The crisis was significant both in terms of its magnitude and its scope, as it spread to other economies. It became a global crisis when it reached the Russian and Brazilian economies and even crossed regional borders with the Russian contagion affecting Brazil (Johnson et al., 2000).

By the end of the century, countries created the financial market infrastructure worldwide, as well as legislation in response to the crises outbreaks. However, the legal actions appeared to be not enough with a lot of blind spots that were successfully used by the managers of the corporations.

The corporate scandals of the 2000s in the USA and Europe have shown the need for more detailed requirements and stricter control over managers of the corporations, with independent bodies chosen for the control. The Sarbanes-Oxley Act of 2002 is an example of the answer to the crisis. Federal law created broad audit practice and financial regulations for public companies to protect shareholders, employees, and the public from accounting errors and fraudulent financial practices. In the USA, the Sarbanes-Oxley Act can be interpreted as a near-perfect mirror of Enron's demise.

The financial crisis of 2008, or the global financial crisis, underscored the importance of not solely relying on government participation in corporate governance development. While governmental bodies can impose mandatory rules to ensure the basic safety of the financial sector, they cannot guarantee the prompt integration of best practices or the proper motivation of participants. Companies may adopt a checkoff approach, where they comply with the rules merely for formal purposes, but managers may still exploit blind spots for self-enrichment. Human ingenuity knows no limits, and therefore, voluntary compliance is crucial.

Aside from stricter rules imposed by governments, major stock exchanges around the world also implemented corporate governance listing requirements to prevent similar problems from happening in the future. In 2008, the OECD initiated an ambitious action plan to develop recommendations aimed at improving areas such as remuneration, risk management, board practices, and shareholder rights (OECD, 2022).

The short-termism in management was identified as one of the key factors contributing to the 2008 financial crisis. Researchers highlighted this structural factor together with an all-powerful CEO (the duality factor and related board independence issues), weak system of management control, weak code of ethics, and opaque disclosures (Grove & Victoravich, 2012). However, it should be noted that short-termism is deeply ingrained in the structure of capitalism itself, which raises questions about the effectiveness of capitalism in its current form.

The deeply grounded issues of the notion of corporate governance and capitalism in general became of particular relevance in the last *crises of the 2020s*: COVID-related crisis in 2020 and politically related crisis in 2022.

The previous crisis were mainly endogenous in nature, as they took their roots from corporate misbehaviors such as accounting fraud, excessive risk-taking, and top management greed and were enabled by poor regulation and the lack of enforcement. Consequently, they stimulated changes in corporate law and governance codes in several countries, aimed at strengthening corporate controls and increasing sanctions in case of misbehaviors (Zattoni & Pugliese, 2021).

On the contrary, the crises induced by COVID-19 or political affairs of 2022 originated neither from corporates' misbehaviors nor by deficiencies in control systems. Crisis of 2008 already put the questions about the ability of the corporate governance to meet new challenges and to change from inside instead of complying to the regulation. The 2020s made it clear that regulatory interventions will be ineffective, as it was in previous crises.

10.2 The 2020s Challenges

Challenges of the 2020s are different from previous crises: they are exogenous unlike previous crisis, they have not solely economic nature, and they impact all spheres of life. In the current complex and interconnected world, the corporate governance frameworks that corporations have built may not be optimal for today's challenges. It is important for corporate governance to evolve and respond to changing environments, beyond just addressing the traditional management-shareholder problem.

The year 2020 has been marked by the *COVID pandemic* and related crisis.

The pandemic itself was first documented in December 2019, with the immediate and long-run impacts on social and economic spheres of life. The COVID-19 pandemic damages had an enormous scope with the fast worldwide exposure of the virus and adverse impacts on each person's everyday life with no exclusions. Along with the instant health impacts, the virus and prevention strategies with quarantine regimes have disturbed the social life and financial security as well as economies at large.

The *social impact* of the virus is the most visible one and the most detrimental. There is a whole list of negative influences:

- Psychological distress due to the rising sickness among family and friends and due to the isolation periods.
- Heightened communication inequalities.
- Food insecurity as a response to the supply chain disruption.
- Diminished access to healthcare in the periods of high rate of infected people.
- Education quality downfall with unpreparedness to the online options.
- Gender-based violence in households (Chu et al., 2020).

The COVID-19 pandemic adverse impacts were most harmful for those living in poverty, the elderly, people with disabilities, children, and indigenous people. For instance, homeless individuals are particularly vulnerable to the threat of the virus since they may not have access to a secure place to stay. Additionally, those without access to running water, refugees, migrants, or people who have been forcibly removed from their homes are also at risk of suffering disproportionately from the pandemic and its aftermath. These individuals may face limitations on their ability to move around or find job opportunities and may even experience increased xenophobia. As a result, the COVID-19 pandemic has the potential to increase inequality, exclusion, discrimination, and global unemployment in the medium and long term (UN, 2021).

In addition to the substantial burden on healthcare systems, COVID-19 has had major *economic consequences* for the affected countries. The COVID-19 pandemic has had a direct impact on revenues due to early fatalities, decreased productivity, and workplace absenteeism. It has also generated a negative supply shock, with manufacturing activity slowing down, as a result of supply chain disruptions and factory closure worldwide. The COVID-19 spread's effects have had a significant influence on international financial markets. The global financial and energy markets sharply plummeted as the number of instances began to rise: S&P 500 index fell by 29% from January 2020 to the middle of March 2020. FTSE 100, CAC 40, and DAX also have lost 25% of their value during 3 months since the beginning of the year with oil prices falling by more than 65% as of April 24, 2020 (Pak et al., 2020).

Soon after COVID pandemic, the year 2022 has made adjustments to the activities of companies. The *geopolitical crisis of 2022* has manifested itself at the global level: supply chains and international trade have been disrupted, social mobility has been limited, and economies started fragmentation. The crisis has non-economic nature and extensive consequences in all areas of the triple bottom line: economic, social, and environmental.

Still suffering from the effects of the economic shocks of the pandemic, the world economy has been experiencing another round of decline in economic activity. The forecast for world GDP growth was to slow from 6.1% in 2021 to 3.2% in 2022 (IMF, 2022). Countries are expected to face a number of challenges, including the rise in the cost of financial capital, an increase in inflationary pressure, a decrease in business activity and consumer demand due to falling incomes, a decline in investment activity, the reduction of government support measures, and a contraction of the labor market (Department of Economic and Social Affairs, 2022; World Bank, 2022).

A sharp *slowdown in economic growth* implies a slowdown in income per capita, which inevitably leads to serious social upheavals. It is predicted that by 2023, per capita incomes will be below the level of 2019 for two-fifth of developing countries. The decline in economic activity and income levels will have far-reaching social consequences for society, including increased inequality, rising unemployment and poverty rates, and even criminal activity. The rising cost of food, particularly in Africa and Western Asia, is also becoming an urgent issue, with implications for food security (World Bank, 2022). The challenge of achieving an inclusive recovery is further complicated by the rise in inflation, which has a disproportionate impact on low-income households that spend a larger portion of their income on food. The decline in real incomes is particularly pronounced in developing countries (Department of Economic and Social Affairs, 2022).

Under the difficult economic conditions, *environmental and climatic risks* arise as well. Energy and food security issues dominate the political discourse, while climate change issues have receded into the background. Meanwhile, despite global CO_2 emissions dropping by 5% in 2020, they already bounced back to the previous estimations in 2021 (Statista, 2022).

As a consequence of the reduction in funding caused by the geopolitical crisis, programs aimed at implementing energy transition, both by companies and governments, may be deprioritized in favor of anti-crisis measures.

The increase in energy prices, particularly natural gas, may also result in the substitution of natural gas with fossil coal, which is more harmful to the environment.

Rising food prices may lead to a reduction in the use of biofuels or to the clearing of land to increase agricultural land (land use changes contribute significantly to greenhouse gas emissions, accounting for about 10% of total emissions in 2019).

High nickel prices (prices have increased by 50% by March 2022) may also have an adverse impact on the environment, as the prospects for higher profits will stimulate the production of additional nickel due to polluting open-pit mining.

The carbon intensity of transportation is likely to increase significantly if the consumption of biofuels (ethanol, biodiesel, and renewable diesel) decreases during the period of high food prices (biofuels are produced from materials suitable for nutrition). The shortage of materials for batteries of electric cars and increased prices for electric cars are also contributing to the rising carbon intensity of ground transportation (World Bank, 2022).

10.3 Does Sustainable Governance Sustain in the Recent Crises Situations?

Corporate governance should be ready to respond to new challenges: be flexible, take into account diverse factors (economic, political, social, and environmental), be an open structure to take into account stakeholder requests and requirements, and, finally, ensure the survival of the company in a crisis.

Sustainable corporate governance that is based on the stakeholder theory views, unlike traditionally accepted corporate governance, meets new requirements and challenges.

Firstly, such corporate system makes it possible to *fully integrate and coordinate the interests of stakeholders* related to the long-term sustainability of the company, control nonfinancial risks, and develop strategic directions of the further development. All this can help companies in the current economic conditions to establish long-term time horizons while making corporate decisions, resist short-term market pressure, increase stability and long-term productivity during the crisis, and reduce the adverse impact.

Sustainable corporate governance is ready for economic and political drivers, as *it is dynamic*: decisions within such system are made not individually within the framework of solely profit maximization goal but jointly with stakeholders in order to create value, measurable both in financial terms and in the form of reputation, brand, innovation potential, consumer loyalty, etc.

Finally, sustainable corporate governance is able to manage not only the financial aspect of activities but is *able to work with a triple-bottom line performance*, which has become fundamentally relevant in recent decades. Sustainable corporate governance covers all relevant spheres of concern in the twenty-first century: climate change, biodiversity, efficient use of resources, waste reduction, deforestation, quality of jobs, decent wages, and respect for human rights.

Recent crises have highlighted the need for corporate governance to transform into sustainable governance by including stakeholders in the decision-making process, taking into account the dynamism of the market, and considering the triple bottom line. The recent changes in the economy have made economic agents more interdependent, increased stakeholder tensions, and made resources scarcer.

Such conditions call for higher degrees of cooperativeness, modifying the competitiveness ethics and postulating that company needs to create value differently: the value should be generated and can be generated only together with stakeholders and for a wide range of stakeholders (Baghiu, 2020; Cardoni et al., 2020; World Economic Forum, 2020). The mutually supportive reaction of the local communities and companies to the external environmental threats may not only allow surviving but also leading to win-win outcomes for all the participants.

10.4 How Should Sustainable Governance Look like in Crisis?

The crises triggered by the COVID-19 and geopolitical conflicts differ significantly from recent systemic economic crises (e.g., Asian Financial Crisis in 1997; the dot. com bubble in 2000; the Global Financial Crisis in 2008). The latter were endogenous in nature, as they originated from corporate misbehaviors, such as accounting fraud, excessive risk-taking, and top management greed, and were facilitated by poor regulation and a lack of enforcement. As a result, crises triggered changes in corporate law and governance codes worldwide, aimed at strengthening corporate controls and increasing sanctions in cases of misbehavior (Zattoni & Pugliese, 2021).

The challenges of 2020 are, on the contrary, exogenous and not purely economic in nature, with massive social and environmental consequences. Therefore, changes in laws or regulations alone are not going to be as effective as before. Corporate governance needs to be changed from within, at its core, in order to meet the various social, political, and environmental obstacles, be open to communication with a wider list of interested parties, and be ready for change.

10.4.1 Agency Theory Metrics to Keep

The initial aim of corporate governance was to control the execution of shareholders' rights and navigate principal-agent conflicts, which continues to be one of the most urgent issues during difficult times. Firms are facing threats to their survival and require equity injections to survive the crisis, leading to mergers and acquisitions, state interventions, and other corporate restructuring, which may cause conflicts between controlling and minority shareholders.

The control function of corporate governance should be implemented and considered a basic condition for overcoming the crisis. Some researchers have empirically proven that in the weeks following the outbreak of the COVID-19 pandemic, companies with tight controls experienced lower declines in stock prices, while firms owned by hedge funds experienced larger drops (Zattoni & Pugliese, 2021). To this end, standard elements of corporate governance should be in place, such as an audit committee, independent directors, no CEO duality, managerial compensation rules, and compliance.

Since sustainable governance and stakeholder theory elements start to play larger role in the company's success, the particular elements have shown the potential to be an anti-crisis remedy.

10.4.2 Strategy and Risk Management

Strategic approach is the only way to lead a company toward sustainable growth and outperforming peers. By taking a strategic approach, companies can overcome the short-termism embedded in the capitalist structure and the agency theory of corporate governance. This approach allows companies to navigate stakeholder requests and build long-term partnerships with win-win outcomes.

However, during times of crisis, there is a high risk that directors may shift their focus from a strategic level to an operational one. While this may be acceptable in the short term (a few months or even days) in order to overcome unexpected obstacles and implement critical changes for survival, an operational focus and short-term thinking are not sustainable in the long run. In fact, the long-term survival of the company will depend on its ability to transition from an ad hoc, reactive operational mode to a proactive planning and strategizing mode.

To maintain a long-term orientation, one potential solution could be to link board remuneration and CEO compensation not only to firm outcomes but also to the board's assessment of behaviors. Risk management is a crucial strategic tool in its own right, and it is essential to apply it to recent challenges. The pandemic and geopolitical crisis should be integrated into the risk assessment process and reported in the annual report, along with other risks that the company faces. By incorporating the management of these recent but significant risks into the highest level of governance, all business functions and elements can be considered in their interconnection, allowing for the assessment of the consequences of risk realization and the exploration of possibilities for minimizing associated concerns.

Example 10.1 Geopolitical Risk Management
ASML Holding N.V. represents a notable example of how a company manages geopolitical risks at the highest level.

ASML is a leading provider of lithographic equipment for the microelectronic industry, producing equipment for the manufacturing of memory chips, flash memory, microprocessors, and more. The company operates in multiple countries, with over 60 service points located in 16 countries. Its headquarter is based in the Netherlands.

The company has large share of the sales in China. In 2021, the total net sales in China amounted to 2740.8 million euros (16.5%). In 2020, net sales in China amounted to 2324.4 million euros (16.6%), and in 2019, total net sales in China amounted to 1377.7 million euros (11.7%). China is the third largest market for ASML, after Taiwan and Korea.

The semiconductor industry has become a battleground in the trade conflict between China and the USA. ASML, being the largest producer of lithographic equipment for the microelectronic industry, has not been immune to this conflict. In 2018, the Trump administration launched a broad campaign to block the sale of Dutch chip manufacturing technology developed by ASML to China. As a result of this campaign, the Dutch government decided not to renew the ASML export license, and equipment worth $150 million was not delivered. Despite these challenges, ASML continues to be a key player in the semiconductor industry, serving customers in various countries and leading innovation in lithographic technology (Alper et al., 2020).

As a result of regulatory demands, the equipment was not delivered to the customer, resulting in economic and reputational costs for ASML. Like many European companies with a substantial share of sales in China, ASML considers geopolitical tensions as threats and risks in the SWOT analysis of its annual reports. Additionally, the company discloses its approach to managing geopolitical risks (ASML, 2022):

> Geopolitical tensions also result in movement restrictions of the employees across countries. Protectionism and bureaucracy are increasing, as well as restrictions impacting international knowledge workers from certain countries (e.g., restricted technology access, visa/travel restrictions). We aim to serve and support all our customers around the world to the best of our ability while being compliant with laws and regulations set by the jurisdictions where we operate. Risk exposure with regard to political tensions, protectionism, and restriction remains high in 2021.

The following are company responses to the risk in several ways (ASML, 2022): monitors geopolitical developments, applies for export licenses as

(continued)

Example 10.1 (continued)
required, complies with (existing and new) regulations, and collaborates with peers in global advocacy.

The geopolitical conflict continues to evolve, while company continues to work with political risks at the highest level: "... the year [2020] also emerged as one with deteriorating geopolitical relationships. The semiconductor industry is one of the battlefields in a trade conflict between the world's superpowers. ASML's position has always been clear: that while the company always operates within applicable rules and regulations, the world's electronics ecosystem is best served with a global and open market, where all players can collaborate and compete on equal terms" (ASML, 2021).

Through a strategic approach to risk management, ASML Holding N.V. has successfully increased its business in China in recent years and continues to expect growth despite geopolitical risks, fulfilling its obligations to key stakeholders including shareholders, owners, employees, and consumers. This success can be attributed to several factors such as a rigorous and consistent risk management approach, transparent communication of the CEO's position on the issue, long-term strategic actions taken to address issues, and ongoing engagement with stakeholders.

10.4.3 Board Composition and Stakeholder Engagement

Although the control function of the board remains an essential rule, in recent years, there has been a shift in the focus of boards from strict control to consultation.

Boards need to act as an information channel for the company and serve a supportive role by assisting management in navigating through the severe consequences of the crisis. To achieve this, directors must possess a high level of expertise and experience, excellent communication skills, and the ability to quickly facilitate the transfer of knowledge to the company.

During times of turbulence, it is advisable for board meetings to be more frequent and flexible in their duration. Certain experts suggest that the format of the meetings also plays a significant role. For instance, the abrupt shift from in-person meetings to virtual ones may impact interactions with potential risks in terms of process inefficiencies and reduced opportunities for cognitive conflicts and constructive dissent (Zattoni & Pugliese, 2021).

In the face of recent crises with far-reaching social consequences, the need for companies to maintain an open attitude toward the world outside has become crucial. It is important for companies to establish rapid and continuous communication with stakeholders and be transparent about the crisis's impacts. For instance, the CEO or a board representative may make an official statement about the company's position and agenda concerning the changes. Such statements may be included in the annual report to present the opinion in a trusted and verified format to a broader audience.

Alternatively, the same declaration can be made through the company website or social media to ensure timely communication with the public.

Stakeholder theory posits that a company's value is cocreated with and for its stakeholders. The COVID-19 pandemic has demonstrated that maintaining strong relationships with customers, employees, suppliers, and contractors can increase a company's chances of weathering the crisis. Therefore, supporting and nurturing stakeholders can result in a smoother recovery or, at the very least, mitigate the damaging effects of shutdowns, supply chain disruptions, and other external shocks. Companies can announce initiatives to support stakeholders during times of instability, such as local community support programs or comprehensive programs to address unemployment. These initiatives can be carried out in partnership with other organizations to reduce resources needed for the projects and increase their visibility and status.

Example 10.2 Helping Stakeholders in Difficult Times

Airbnb, Inc.—an American company, an online platform for placement and search of housing for rent around the world—has been helping people in times of crisis through the Open Homes initiative for over 8 years.

The Airbnb initiative began in 2012 when Hurricane Sandy hit New York City, and one of the hosts suggested accommodating people who had to evacuate. The initiative gained traction, and in the same year, more than 1000 hosts across the USA joined the program. In 2013, Airbnb launched a disaster response tool that allows hosts worldwide to offer their homes for free during disasters. In 2015, Airbnb mobilized for the Nepal earthquake, and hosts welcomed their first group of relief workers, including volunteers from All Hands and Hearts.

By September 2017, the Open Homes initiative had helped over 20,000 people and had gained significant recognition from the community. As a result, Airbnb officially launched the initiative to expand its organizational status, allowing for more participants and formats of participation. In 2019, Airbnb took another step forward by introducing a donations platform that allows hosts to donate a percentage of their earnings to Open Homes nonprofit partners. This enabled them to help fund stays for those who are in need.

Amidst the COVID pandemic, Airbnb launched a program enabling hosts to offer housing to healthcare workers and first responders at the forefront of the crisis. The company also established Airbnb.org, an independent nonprofit organization that connects hosts and partners to provide housing and resources during times of crisis.

Collaborating with nonprofit organizations such as HIAS and the International Organization for Migration, Airbnb.org offers housing to refugees and other individuals in need. Airbnb has been involved in helping refugees since the beginning of the conflict in Ukraine, and Airbnb.org has been funding short-term housing for up to 100,000 people who are fleeing the country, as of year 2022 (*Airbnb.Org—About*, 2021).

10.4.4 Anti-Corruption and Business Ethics

It is important to incorporate changes into the value system and communicate the understanding of the situation to all employees and stakeholders.

Many individuals were unsure how to manage their health and daily lives during the pandemic, given the new conditions of self-isolation, restrictions, and limitations. Therefore, it was vital to provide individuals with information about safety, ethical guidance, and support when they were in emotional distress. Empirical research already supports the role of corporate culture in turbulent times. According to Li et al., companies with a strong corporate culture outperform their peers without strong corporate ties (Li et al., 2021).

Example 10.3 Support of the Emotional Well-Being
Merkle, Inc., a performance marketing agency, presents an example of how to help people in overcoming the dizzying pace of COVID-19 news. For this, the company has utilized its competencies in consulting, database marketing, customer insight enablement, and digital media services on a global scale.

Merkle created a platform called COVID-19 roundup that pulls together key news and emerging trends and pairs it with Merkle interpretations and insights, navigating people and helping to prioritize the information from wide list of mass media sources. The aggregated information helps in understanding the COVID restrictions in different countries and regions, industry-specific changes for companies and their employees, and recommendations on personal health and safety measures. For business partners, Merkle created the COVID-19 interactive dashboards that combine county-level medical and healthcare coverage data with Google Mobility and consumer spend data to help guide an appropriate engagement strategy for the organization (MERKLE, 2021).

During the COVID-19 pandemic, information was dispersed, voluminous, unstructured, and at times controversial. This created fear, panic, and increased stress levels, disrupting the everyday lives of people. Merkle, a performance marketing agency, utilized their expertise in consulting, database marketing, customer insight enablement, and digital media services on a global scale to present filtered, correctly interpreted, and nicely presented data. This tool was instrumental in combating disinformation and noise while helping stakeholders stay up-to-date with minimal stress. In turn, the company maintained communication with its customers and counterparts, finding common ground in understanding the challenges ahead.

10.4.5 Transparency

Following stakeholder theory, reporting on the effects of a crisis can be a valuable tool for companies to maintain transparency and communication with stakeholders. Engaging with stakeholders and providing information on adverse effects are factors that companies should consider to involve stakeholders in the mutual process of overcoming the downturn. Including such information in a timely manner demonstrates the agility of the corporate structure and transparency to stakeholders. Furthermore, scholars argue that disclosing information should not be limited to powerful stakeholders but should also be extended to vulnerable ones who are likely to be most affected by the information (Zharfpeykan & Ng, 2021).

If a company supports stakeholders in an unstable situation and promptly reports on key changes, it can help the firm overcome the crisis with the strong support of solvent consumers, loyal suppliers, and supportive local communities.

Example 10.4 Reporting about Crisis

Relevant disclosures and reporting play a crucial role in the effective communication of a company's performance, policies, and practices to its stakeholders. While various modes of reporting exist, the annual report, along with its variants such as the integrated report, sustainability report, ESG report, etc., serves as the most dependable source of information. These public documents are signed by the CEO and/or Chair of the Board and are verified by the internal auditor. Furthermore, third-party auditing of such reports is a common practice.

The reports are typically published annually, with a lag of several months, which means that key events from the following year are often omitted from the document. It would be advisable to include an explanation of the most relevant events in the report, even if they occur after the reporting date. Events such as the COVID-19 pandemic or geopolitical crises can have a significant impact on all stakeholders and may alter the significance of past events.

The European Bank for Reconstruction and Development presents an example of publishing the information about recent changes in the report. The 2019 sustainability report of EBRB includes a separate block on COVID impact, where EBRD describes Solidarity Package with its investments into recovery of economics and measures to combat COVID at EBRD as an organization, such as workplace health and safety, information policy in relation to employees, and other stakeholder communication. Promoting environmentally sound and sustainable development and appropriate procurement practices, the Bank is expected to be on the forefront of battling the COVID pandemic and related socioeconomic crisis (EBRD, 2020).

Another company that has been significantly impacted by the COVID pandemic and has been required to undertake unprecedented measures to

(continued)

Example 10.4 (continued)
overcome it is Pfizer, an American multinational pharmaceutical and biotechnology corporation. Pfizer has included the pandemic issues in the risk section of its annual report, thereby integrating the issue into the core structure of the report. This approach ensures that the impact of the pandemic is accurately reflected and accounted for in the company's reporting, providing stakeholders with a comprehensive understanding of the risks and challenges faced by the company in the current business landscape:

"Public health epidemics or outbreaks could adversely impact our business. In December 2019, a novel strain of coronavirus (COVID-19) emerged in Wuhan, Hubei Province, China. While initially the outbreak was largely concentrated in China and caused significant disruptions to its economy, it has now spread to several other countries and infections have been reported globally. The extent to which the coronavirus impacts our operations will depend on future developments, which are highly uncertain and cannot be predicted with confidence, including the duration of the outbreak, new information which may emerge concerning the severity of the coronavirus, and the actions to contain the coronavirus or treat its impact, among others. In particular, the continued spread of the coronavirus globally could adversely impact our operations, including, among others, our manufacturing and supply chain, sales and marketing, and clinical trial operations, and could have an adverse impact on our business and our financial results" (Pfizer Inc., 2021).

Example 10.5 CEO Communication
The CEO of a company may convey the main ideas through a personal letter in the annual report or the company's website or even via social media. Herbert Diess, who served as the CEO of Volkswagen Group from 2018 to 2022, actively used social media platforms to communicate with the public. He wrote numerous posts highlighting innovations within the company and also shared his thoughts on global events, including those related to the COVID-19 pandemic and the geopolitical crisis of 2022.

Acknowledgments The Sustainable Corporate Governance Rating organizers played valuable role in a discussion regarding the selection of key sustainable governance metrics for companies that were significantly impacted by the crises of the 2020s (Sustainable Corporate Governance Rating experts, 2021).

References

Airbnb.org. (2021). Airbnb.Org. Retrieved 12 September, 2022, from https://airbnb.org/about

Alper, A., Sterling, T., & Nellis, S. (2020). *Trump administration pressed Dutch hard to cancel China chip-equipment sale: Sources*. Reuters. Retrieved 11 September 2022, from https://www.reuters.com/article/us-asml-holding-usa-china-insight/trump-administration-pressed-dutch-hard-to-cancel-china-chip-equipment-sale-sources-idUSKBN1Z50HN

ASML. (2021). *ASML annual report 2020*. Retrieved May 06, 2023, from https://www.asml.com/en/investors/annual-report/2020

ASML. (2022). *Annual report 2021 | ASML*. Retrieved May 06, 2023, from https://www.asml.com/en/investors/annual-report/2021

Baghiu, M. C. (2020). Anallysis of business model innovation in post-COVID economy: Deterninants for success. *Journal of Public Administration, Finance and Law, 17*, 7–24.

Berle, A. A., & Means, G. (1932). *The modern corporation and private property*. Commerce Clearing House. https://edisciplinas.usp.br/pluginfile.php/106085/mod_resource/content/1/DCO0318_Aula_0_-_Berle__Means.pdf.

Cardoni, A., Kiseleva, E., & Taticchi, P. (2020). In search of sustainable value: A structured literature review. *Sustainability, 12*(2), 615. https://doi.org/10.3390/su12020615

Chu, I. Y.-H., Alam, P., Larson, H. J., & Lin, L. (2020). Social consequences of mass quarantine during epidemics: A systematic review with implications for the COVID-19 response. *Journal of Travel Medicine, 27*(7), taaa192. https://doi.org/10.1093/jtm/taaa192

Demsetz, H., & Lehn, K. (1985). The structure of corporate ownership: Causes and consequences. *Journal of Political Economy, 93*(6) 1155–1177. https://doi.org/10.1086/261354

Department of Economic and Social Affairs. (2022, June). *World economic situation and prospects*. Retrieved September 07, 2022, from https://www.un.org/development/desa/dpad/publication/world-economic-situation-and-prospects-june-2022-briefing-no-161/

EBRD. (2020). *Sustainability report 2019*. Retrieved September 07, 2022, from https://2019.sr-ebrd.com/

Grove, H., & Victoravich, L. (2012). Corporate governance implications from the 2008 financial crisis. *Journal of Governance and Regulation, 1*, 68. https://doi.org/10.22495/jgr_v1_i1_p7

IMF. (2022, July). *World economic outlook update, July 2022: Gloomy and more uncertain*. IMF. Retrieved September 07, 2022, from https://www.imf.org/en/Publications/WEO/Issues/2022/07/26/world-economic-outlook-update-july-2022

Johnson, S., Boone, P., Breach, A., & Friedman, E. (2000). Corporate governance in the Asian financial crisis. *Journal of Financial Economics, 58*(1), 141–186. https://doi.org/10.1016/S0304-405X(00)00069-6

Li, K., Liu, X., Mai, F., & Zhang, T. (2021). The role of corporate culture in bad times: Evidence from the COVID-19 pandemic. *Journal of Financial and Quantitative Analysis, 56*(7), 2545–2583. https://doi.org/10.1017/S0022109021000326

Merkle. (2021). *COVID-19 resources*. Retrieved 12 September 2022, from https://www.merkle.com/thought-leadership/covid-19-resources

OECD. (2022). *Corporate governance and the financial crisis*. Retrieved September 07, 2022, from https://www.oecd.org/daf/ca/corporategovernanceprinciples/corporategovernanceandthefinancialcrisis.htm

Pak, A., Adegboye, O. A., Adekunle, A. I., Rahman, K. M., McBryde, E. S., & Eisen, D. P. (2020). Economic consequences of the COVID-19 outbreak: The need for epidemic preparedness. *Frontiers in Public Health, 8*, 241. https://www.frontiersin.org/articles/10.3389/fpubh.2020.00241

Pfizer Inc. (2021). *Investor relations—financials—annual reports*. Retrieved 16 September 2022, from https://investors.pfizer.com/Investors/Financials/Annual-Reports/default.aspx

Statista. (2022). *Annual CO_2 emissions worldwide 1940–2020*. Retrieved 16 September 2022, from https://www.statista.com/statistics/276629/global-co2-emissions/

Sustainable Corporate Governance Rating experts. (2021). *Sustainable Corporate Governance Rating website*. Retrieved 31 January 2023, from http://corptransparency.ru/partners

UN. (2021). *Everyone included: Social impact of COVID-19*. Retrieved 11 September 2022, from https://www.un.org/development/desa/dspd/everyone-included-covid-19.html

World Bank. (2022). *Global economic prospects, June 2022*. World Bank. Retrieved 11 September 2022, from https://doi.org/10.1596/978-1-4648-1843-1

World Economic Forum. (2020). *Measuring Stakeholder capitalism: Towards common metrics and consistent reporting of sustainable value creation*. Retrieved 11 September 2022, from https://www.weforum.org/reports/measuring-stakeholder-capitalism-towards-common-metrics-and-consistent-reporting-of-sustainable-value-creation/

Zattoni, A., & Pugliese, A. (2021). Corporate governance research in the wake of a systemic crisis: Lessons and opportunities from the COVID-19 pandemic. *Journal of Management Studies, 58*(5), 1405–1410. https://doi.org/10.1111/joms.12693

Zharfpeykan, R., & Ng, F. (2021). COVID-19 and sustainability reporting: What are the roles of reporting frameworks in a crisis? *Pacific Accounting Review, 33*(2), 189–198. https://doi.org/10.1108/PAR-09-2020-0169

Chapter 11
Challenges in Sustainable Corporate Governance Development

11.1 Developed Economies in the Strive for Best Practice Sustainable Corporate Governance

In the context of developed countries, the "cycle of sustainability" involving large companies has been activated since the last decade, and there has been a growing interaction between the main actors of the socioeconomic system (Fig. 11.1).

The major issues affecting the effective implementation of sustainability practices, such as institutional requirements, ESG topics, reporting evolution, and assurance standards, have influenced the regulatory and professional behavior at institutional and managerial level, impacting companies' orientation in the attempt to conform their strategic and operational processes to the new ESG paradigms.

In such interaction, with specific reference to the theme of sustainable governance, it is possible to identify the following challenges.

The first challenge concerns a *real integration of sustainable governance with corporate strategy and business models*, avoiding behaviors and initiatives conditioned by a greenwashing purpose.

Greenwashing is considered a threat to accurate ESG information that can affect also the G (governance) components. As suggested by Bowen and Aragon-Correa (2014) and Du (2014), greenwashing is a deliberate information disclosure decision initiated by firms that may be beneficial to firms but costly to society. Literature identifies three types of greenwashing associated with the following:

- Manipulated disclosure to boost company valuation (Marquis et al., 2016), occurring when firms adopt "greenwashing strategy" to obscure their poor environmental performance by disclosing large quantities of environmental data to mislead their stakeholders.
- Selective disclosure to mislead investors, related to the deliberate choice to report only positive environmental information but hiding negative information (Lyon & Montgomery, 2013; Marquis et al., 2016). In other circumstances, Kirk and Vincent (2014) document how companies disclose private information to a

A. Cardoni, E. Kiseleva, *Sustainable Governance*, CSR, Sustainability, Ethics & Governance, https://doi.org/10.1007/978-3-031-37492-0_11

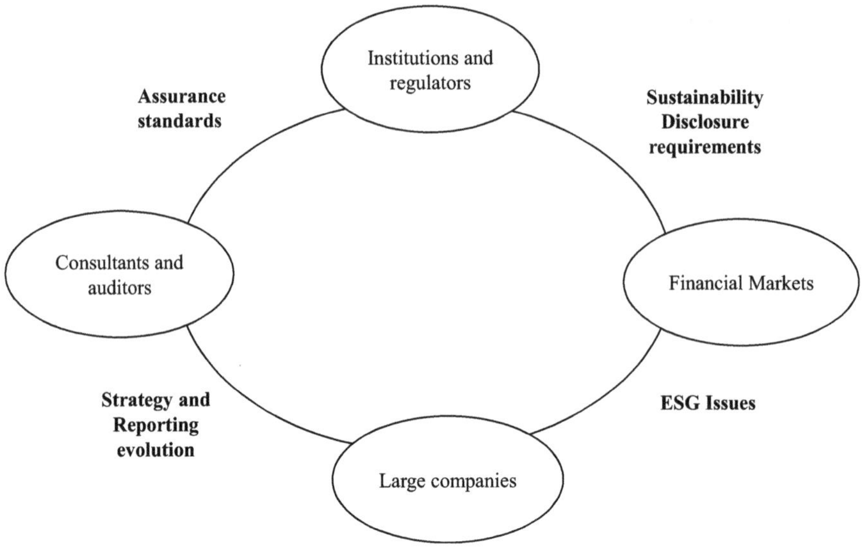

Fig. 11.1 The "cycle of sustainability" for large companies in developed economies. Source: constructed by the author

selected group of investors only. Consequently, these firms are able to create a false impression to mislead the public about their actual environmental performance.
- Product-level greenwashing (Cho & Baskin, 2018), in all the cases when the manipulated or selective information is referring to products rather than firm-level perspective.

Recent studies demonstrate how governance can have a strong impact on the containment of such practices (Yu et al., 2020). Particularly, there are several characteristics of the governance that are helpful to this extent, like the share of institutional investors and the share of independent directors, since it can increase the board's capability for monitoring (Yu et al., 2020).

A better quality of governance certainly favors a real integration of sustainability with the strategy, having positive effects on performance outcomes in several perspectives: financial performance, image and reputation, stakeholder perceptions, and cultural change (Hristov et al., 2022). However, the impact of sustainable governance on performance is not always visible, and this is a major challenge for companies. Many researches mentioned that a large part of the decision is still driven by economic expectations in terms of cost reduction and revenue growth (Bhattacharya & Sen, 2004). Managers continue to be too focused on short-term financial results, a critical issue that hinders the integration process, calling for a cultural change to achieve an integrated view of the corporate system. Specific qualitative and quantitative measures are thus needed to evaluate the performance of the sustainability strategy adopted. Recent research (Hristov et al., 2022) has proposed a sustainable KPI system as a tool to support the integration process in view of long-term advantages, translating the firm's strategic objectives into a set of measures leading to sustainable value creation. For each dimension, a sustainability plan needs to be prepared based on a multidimensional scorecard of the strategic goals, KPIs, measures, processes, targets, and analyses. This approach is essential to design a structured system that allows board of directors and managers to implement sustainable governance based on the fundamental drivers of sustainability development.

The second challenge of sustainable governance in developed countries refers to *the quality of sustainability disclosure*. Sustainable governance and corporate disclosure are characterized by a mutual interaction. As discussed earlier, research shows how sustainable governance impacts the quality of reporting (Erin et al., 2021), but at the same time a good disclosure of governance components constitutes one of the most relevant pillar to evaluate the corporate transparency. On the first type of interaction (impact of sustainable governance on disclosure), research has broadly demonstrated the positive relationship of various governance characteristics, such as the following:

- Board size (Dienes et al., 2016).
- Independence (Odriozola & Baraibar-Diez, 2017).
- Expertise (Michelon & Parbonetti, 2012).
- Audit committee (Arumona et al., 2019).

The challenge of further development is linked to evolution of governance toward greater integration with strategy, as highlighted above. In this perspective, a significant role is played by accounting (Rinaldi, 2019). Particularly, accounting can support the strategic integration into sustainable governance and could be used in three different approaches, as demonstrated by classifying the recent literature (Rinaldi, 2019):

- Accounting for sustainable governance as a "management program" that focuses on the role of accounting within a space of management and measurement frameworks, analyzing the role of ESG on disclosure practices and internal processes more broadly.
- Accounting for sustainable governance as a "financial program" that explores how accounting for sustainability can influence responsible investing and the characters of financial markets, with particular reference to the impact of selected ESG characteristics on firm value.
- Accounting for sustainability governance as a "social and environmental program," referring to the calculative mechanisms put in place to generate social and environmental outcomes, acting as a stimulus to translate sustainability ambitions into socially and environmentally responsible behavior.

On the second perspective (disclosure of sustainable governance components), one of the most relevant challenges for ESG data disclosure, including governance, is that sustainability reports are unaudited, and there is lack of standardization in disclosure rules and no global governing body to ensure the accuracy of reported

ESG information. The standardization of sustainability reporting frameworks is one of the major issues discussed at worldwide level, with the aim to find the right balance between the qualitative adaptability for the companies to communicate their material topics to relevant stakeholder and the common metrics to support the financial markets investment decisions (Adams & Abhayawansab, 2022). Considering the complex interaction between the relevant key players in the "cycle of sustainability" (Fig. 11.1), a possible solution is to implement a "collaborative" governance (Wang et al., 2022). Extending the focus on media, companies, and government, recent studies (Wang et al., 2022) demonstrate that an enhanced synergism of government regulation and media reports can increase the cost and loss of poor disclosure and greenwashing, promoting the internal supervision to play a more structured approach of sustainable governance.

The third challenge, primarily affecting the European context, concerns the *regulatory profile* of directors' liability. The EU legislator initiated a reform program in 2018 with the formal adoption of the "Action Plan: Financing Sustainable Growth," with the aim to promote the sustainable corporate governance and the reduction of short-termism in capital markets. Consistently with this approach, the EU performed an analytical and consultative work with relevant stakeholders to assess the possible need to clarify the rules according to which directors are expected to act in the company's long-term interest. On this specific topic, in July 2020, the "Study on Directors' Duties and Sustainable Corporate Governance," prepared by EY (2020), identified the core problem as a "trend for publicly listed companies within the EU to focus on short-term benefits of shareholders rather than on the long-term interests of the company." In light of this issue, the study called for EU intervention to tackle short-termism and ensure a level playing field for European companies in pursuing three specific objectives:

1. Strengthening the role of directors in pursuing their company's long-term interest.
2. Improving directors' accountability toward integrating sustainability into corporate decision-making.
3. Promoting corporate governance practices that contribute to corporate sustainability.

On February 2022, the European Parliament issued the proposal for a Directive on Corporate Sustainability Due Diligence, mainly with the following aim:

- Better integrate risk management and mitigation processes of human rights and environmental risks and impacts.
- Avoid fragmentation of due diligence requirements in the single market and create legal certainty.
- Increase corporate accountability for adverse impacts.
- Improve access to remedies for those affected by adverse human rights and environmental impacts of corporate behavior.
- Become a horizontal instrument focusing on business processes, applying also to the value chain.

In terms of scope of application, the directive is applicable for the company with more than 500 employees on average and a net worldwide turnover of more than EUR 150 million. The size of a company may be even smaller (more than 250 employees on average and a net worldwide turnover of more than EUR 40 million), provided that at least 50% of this net turnover was generated in one or more of some environmentally sensitive sectors (e.g., textiles, agriculture, forestry, wood, food, and beverages; extraction of mineral resources, etc.).

Under the directive, companies will be required to identify and, where necessary, prevent, end, or mitigate adverse impacts of their activities on human rights and the environment. This corporate due diligence duty applies to companies' own operations, their subsidiaries, and their value chains. The directive also requires large companies to have a plan to ensure that their strategy is compatible with the goals of the Paris Climate Agreement. Furthermore, the duties of company directors are clarified to ensure that their decisions consider human rights, climate change, and the environment.

The European approach has been criticized for being excessively rigid, imposing a sustainable governance model on a legislative basis, and presenting various complexities on juridical implementation (Ferrarini et al., 2021). Particularly, according to the authors (Ferrarini et al., 2021), it has not been adequately considered the positive role of "soft law" for implementing sustainable governance and the broader context of EU regulations.

In terms of soft law, international organizations and standard setters with governance codes, such as the IMF, the OECD, the World Bank, and the United Nations, already contribute to establish norms of corporate behavior to reduce the negative impact of corporate activities on third parties, being characterized by a good flexibility and international standardization.

Regarding EU regulations, several reforms have been adopted by the EU legislature in recent years, such as the Corporate Sustainability Reporting Directive, the Taxonomy Regulation, the Sustainable Finance Disclosure Regulation, and the Shareholder Rights Directive II. These frameworks address corporate short-termism and aim to promote sustainability in the governance of firms, offering a more promising prospect for sustainable governance than the reform of directors' duties.

This is indeed a significant change in the regulatory systems of European countries, even though the Directive has the potential to influence policymaking and business practices well beyond the EU. Once defined in its definitive formulation, the Directive should be integrated with the national legislative framework, and there is some disagreement among stakeholders about how the proposal should be implemented in practice. Certainly, the Directive's implementation could have a strong impact on sustainable governance at an international level, directly affecting large companies but also the smaller companies of the supply chain. The definitive elaboration and integration on the legislative systems of the European countries certainly represent significant challenges to be faced in the near future. The Directive will require European large companies to undertake more in-depth due diligence on their risks and strategies for meeting their sustainability objectives, including the need for possible culture change (Hobbs, 2023).

11.2 Trends for Developing Economies in Sustainable Governance Adaptation and Development

The development of the ESG agenda in developing countries, and sustainable corporate governance in particular, is still in its early stages compared to developed economies. However, as highlighted in Chap. 8, the significance and importance of implementing sustainable governance is only growing. Regulators, exchanges, investors, media, and consumers all expect businesses to play a role in securing a safe and decent future while also being transparent about their impact on the environment and society.

In contrast to developed economies, where sustainable corporate governance and ESG implementation are already established practices, developing markets often rely more heavily on *government intervention* to drive progress in this area. The motivation to develop a sustainable development agenda is primarily driven by the state's aim to introduce basic practices and standards to facilitate further interaction between national and supranational institutions. Such approach is twofold: from the one hand, centralized approach helps to develop unified and widespread sustainable corporate governance practice. On the other hand, there is a risk to receive low productivity of such system: standardized ESG management may increase costs for the company without a contribution to a greater economic performance (Cardoni et al., 2020; Ioannou & Serafeim, 2019; Porter & Kramer, 2006). Due to the strong influence of the state in business, corporate mechanisms can resemble those of public management. As a result, corporate governance can be transformed and become less significant in the company's operations, sometimes adopting a formalistic approach and imitating corporate governance instead of functioning as an effective system. By using such an approach, companies simply verify the presence of popular ESG tools within their corporate structures and ensure that the "right" procedures and documents are in place but fail to take their own decisions regarding necessary actions.

Together with government participation, *institutionalization* of sustainable governance is increasing at all levels.

The worldwide institutionalization of sustainable development began with the appearance of international organizations such as the Club of Rome and the United Nations, as well as their documents, papers, and events. EU is one of the leaders in institutionalization of sustainable development. In the EU, organizations such as EUROSIF and the European Sustainable Development Network were created, along with standard setters such as GRI, TCFD, CDP, IFRS, and others. The EU has a wealth of regulatory documents, including the Corporate Sustainability Reporting Directive (CSRD) and MIFID for sustainable financial markets, as well as projects like the Sustainable Corporate Governance Initiative by the EU Commission.

Developing countries just have started to put some regulative pressures and create local institutions responsible for adaptation of sustainable development to the local governmental and corporate aims and objectives.

Government is usually responsible for formulating the understanding of the sustainable development directions and the priorities in choosing material issues as well as imposing basic set of instruments in the field corporate governance, ESG and sustainability, obligatory list of topics to disclose by public or large companies, and corporate governance structure regulations. For example, the China Securities Regulatory Commission (CSRC) revised the Code of Corporate Governance for Listed Companies in 2018, which now includes a dedicated chapter on the "stakeholders, environmental protection, and social responsibility" of listed companies in China (China Securities Regulatory Commission, 2019).

Central banks and exchanges play a significant role in the institutionalization of sustainability and sustainable corporate governance. In 2022, several stock exchanges in developing economies issued guidance documents to support companies in their sustainability efforts. For instance, the Johannesburg Stock Exchange in South Africa published the JSE Sustainability Disclosure Guidance, while the Brazilian stock exchange B3 issued the Novo Valor Corporate Sustainability Guide, providing information on how to begin, who to involve, and what to prioritize. The National Stock Exchange of India also released the NSE-SES Integrated Guide to Business Responsibility and Sustainability Report (BRSR) to assist companies in their reporting efforts. Additionally, the Moscow Exchange in Russia published the ESG Best Practice Guide, following in the footsteps to promote sustainable business practices (Sustainable Stock Exchanges Initiative, 2023).

Businesses are increasingly taking steps to institutionalize stakeholder engagement and address their concerns. One notable approach is the establishment of stakeholder committees, which provide a platform for interested parties to participate in decision-making processes and advocate for stakeholder-oriented business models. In some countries, these committees are mandatory for government agencies and companies, while there are ongoing debates about the feasibility of extending such requirements to all public companies. Such initiatives demonstrate the growing recognition among businesses of the need to engage with stakeholders and prioritize their interests in shaping corporate strategies and decisions.

11.3 SME Context

In the context of SMEs, the "sustainability cycle" is characterized by a still unstructured type of interaction between the various actors of the economic-social system, which presents greater uncertainties for the future.

As represented in the figure (Fig. 11.2), the process is currently top-down directed, with the institutions that are stimulating the diffusion of sustainability at all levels, both with the reforms on the credit mechanisms and definition of sustainability reporting requirements for SMEs. At the same time, SMEs already receive strong requests for the adoption of sustainable processes and/or products by their large customers or by other actors in the supply chain.

The great uncertainty regards how SMEs will react to such institutional pressures and how they will be able to activate new behaviors and initiatives at a strategic, governance, and business model level. These changes will necessarily require new

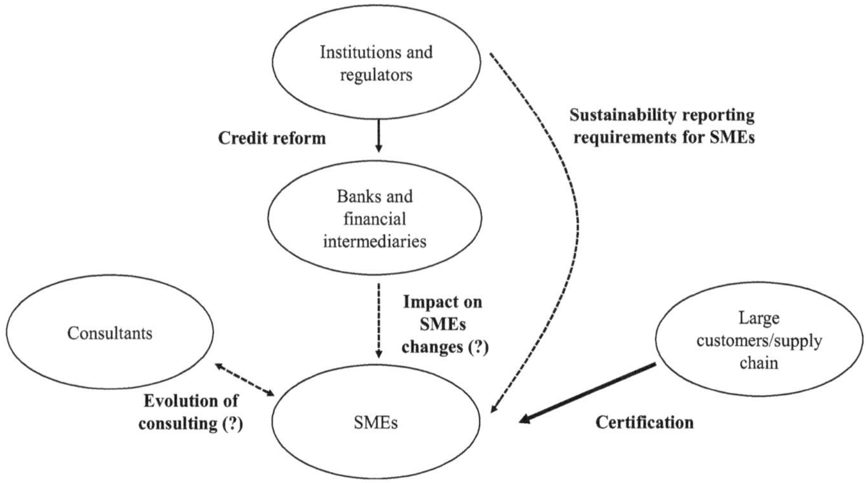

Fig. 11.2 The "cycle of sustainability" for SMEs. Source: constructed by the author

type of relationship with consultants in order to receive professional support not limited to compliance aspects but with a strong strategic and innovative orientation toward sustainability.

This scenario presents relevant challenges for sustainable governance in the context of SMEs.

The first challenge concerns the necessary *evolution of governance models through the implementation of a more structured governance boards and the correct planning of the generational succession.* These challenges are not new for SMEs, having been frequently reported by literature over the last decades (Handley & Molloy, 2022).

Among the most relevant aspects for enhancing corporate governance, the first aspect certainly concerns the correct composition and effective functioning of the boards of directors. In fact, the strategic and managerial processes of SMEs are often governed exclusively by the entrepreneur, without a shared and professional leadership that can bring high-level independent contribution to companies' general management. Many studies demonstrate how the size of board and the presence of an independent member may positively impact on the performance.

The same theoretical and practical awareness has matured with reference to the succession planning (Sgrò et al., 2022). Sustainable governance is, first of all, the ability of the company to maintain conditions of business continuity, not relying on the exclusive set of entrepreneurial capabilities at individual level. As the development cycle of modern economies experienced the launch of many entrepreneurial activities in the 1980s and 1990s, many SMEs will face the challenge of generational succession in the next few years. Finding the right balance between business, family, and strategic needs is an extremely complex challenge, certainly topical to guaranteeing the best conditions of sustainable governance. Considering the peculiarities of the SMEs, characterized by limited financial and managerial resources, high dependence on external events, entrepreneurial individual leadership, and hydiosyncratic attitude toward external directors, these challenges will be strongly demanding but crucial for developing sustainable governance.

It sounds prosaic talking about sustainable governance for SMEs if these innovations, which continue to encounter strong cultural barriers and limitations in managerial resources, are not implemented as necessary preconditions. Even in the policymaker perspective, it should be helpful that incentive mechanisms will be activated to implement these changes and provide SMEs with the necessary support to deal with this radical change in corporate culture.

The second challenge refers to the needed *evolution of SME management control system* to support the sustainable governance. As demonstrated by literature (Lavia Lopez & Hiebl, 2015; Heinicke, 2018), at the beginning of the new millennium, a series of strategic conditions (globalization, innovation, technological advancement, competitiveness) prompted the SMEs to experiment new evolutions in their control models. Despite the organizational efforts, these models have remained very traditional, unable to support innovation and strategy elaboration (Ahmad, 2017). With the advent of the profound discontinuities that the business environment experimented since the financial crisis of 2008, the traditional management control approaches lost their relevance, and SMEs had to mostly rely on entrepreneurial orientation to achieve resilience (Eggers, 2020), interrupting a process of investment of resources and skills on more advanced management accounting tools.

Many SMEs are still managed and governed according to an elementary logic, with a minimum of adequate tools for the strategic governance of the current business environment (Cardoni & Paradisi, 2020). The challenges posed by the actual environment to SMEs require an evolution of management control system that must be integrated with a risk management logic (Arcari, 2018). At the same time, considering the strategic relevance of sustainability, the accounting tools must gradually incorporate the environmental perspective (Castellano & Felden, 2021; Manzaneque-Lizano et al., 2019; Maraghini & Vitale, 2018). As demonstrated by the evolutionary stage approaches, these advancements require a gradual process of evolution on the capacity to elaborate and manage financial and nonfinancial information, which is scarcely implemented in SMEs.

The third challenge relates to *stakeholder communication and disclosure* in their strict connection with sustainable governance advancements. As widely demonstrated, the impact of governance on sustainability disclosure is fundamental also in the SME context, so that the smaller companies will be called to face a strong challenge involving the corporate culture in the next future to comply with institutional requirements. Researches show a complex relationship between the characteristics of SMEs and sustainability disclosures that is thought to be related to several reasons.

Firstly, preparing sustainability disclosure and stakeholder communication requires a complex set of competences involving strategic analysis, organizational knowledge, and a range of related skills (Gjergji et al., 2020). This, in turn, requires the development of specific governance structures, organizational spaces, actors, rules, and professional capital (Hahn & Kühnen, 2013; Lai & Stacchezzini, 2021), which can be challenging to acquire for an SME.

Secondly, the presence of professional managers is limited (Karim et al., 2013), as SMEs are often run by family members (Bank of Italy, 2018) and the reporting process is usually managed by an external consultant (IFAC, 2008) who may tend to adopt a culture of compliance (Calace, 2014).

Thirdly, SME governance is often entrepreneurial, with no active role for the board of directors that conversely is able to play a fundamental role in sustainability disclosure (Cormier & Magnan, 2015; Fasan & Mio, 2016) and strengthen stakeholder engagement, also following the regulatory changes (Aureli et al., 2020; Caputo et al., 2021).

Many organizational and professional conditions must still be established to strengthen the culture of sustainability disclosure in SMEs, to raise awareness, to manage the reporting process with more internal professionalism, and to make external advisors less focused on compliance (Lai & Stacchezzini, 2021). In this regard, we support the European Commission's proposal to develop separate, proportionate standards for SMEs 3 years after they apply to other companies. Only preparing and cultivating in advance the shared set of the competences needed, the mandatory sustainability reports can have a deep impact on business strategies and stakeholder engagement, not remaining only an additional formal burden for SMEs.

Summarizing, many conditions are needed to fully implement these changes, starting with those mentioned above on the evolution of traditional governance issues and management control systems. The simultaneous interaction of these three challenges to be faced together will create the uncertainty about the activation of the "sustainability cycle" and sustainable governance for SMEs.

11.4 Reporting and Analyzing Sustainable Governance Data under Industry 4.0 Challenges

Despite the value relevance of ESG issues, there are several difficult points that limit usage of ESG data in the investment practice. One of the main factors that limit the usage of ESG information is the lack of comparability of ESG data across companies, including sustainable governance data: being one of the most comparable among ESG issues, governance data is still not unified at the same extent as financial data.

In an attempt to achieve comparability of the data, standard setters are moving toward a unified understanding of the metrics. The level of standardization in the sphere of ESG reporting frameworks today is low and provides variability in reporting formats, such as GRI reports, integrated reports, and SASB/CDP/TCFD reporting. At the end of 2021, the International Sustainability Standards Board (ISSB) has published the Exposure Draft IFRS S1 General Requirements for

Disclosure of Sustainability-Related Financial Information. IFRS standards have the potential to become the most widespread because of the united forces of the Value Reporting Foundation, IFRS, SASB, TCFD, and other bodies in the creation of these requirements. However, they are still in draft form and under comments in 2022 and are not ready for widespread adoption (IFRS-Exposure Draft and Comment Letters: General Sustainability-Related Disclosures, 2022). Because of such divergence in sustainability reporting frameworks, the availability of sustainability reports does not guarantee the availability of a particular content or some level of quality of ESG in the company.

Government pressure may push companies to provide data in accordance to the more standardized recommendations. The best example is Non-Financial Reporting Directive (Directive 2014/95/EU) development in EU. The NFRD sets the rules on disclosure of nonfinancial and ESG information relating to the ESG areas (environmental protection, social responsibility and treatment of employees, respect for human rights, anti-corruption and bribery, and diversity on company boards) by certain large EU public companies in their annual reports. In April 2021, the Commission put forward a proposal for a Corporate Sustainability Reporting Directive (CSRD), which has significantly broaden the reporting requirements of the NFRD, making the field of ESG reporting more and more standardized.

Technological tools that appear in the field may largely help in increasing the data comparability. The so-called Fourth Industrial Revolution and Industry 4.0 have disrupted society's understanding of business and the value creation process. Digitalization has been identified as the main driver of change in all sectors of the new economy. In this context, intangible assets, such as patents, knowledge, human resource capabilities, etc., have become the main part of a company's value. The scale of economic value alone is not adequate in measuring the growing contribution of intangible assets.

One of the technological instruments for ESG data is *IXBRL reporting*, which can be applied to ESG metrics as well as to financial metrics. The ESEF Regulation introduces a single electronic reporting format for the annual financial report of issuers whose securities are listed on regulated EU markets. Issuers must prepare and disclose their annual financial statements in xHTML format. Issuers publishing consolidated financial statements in accordance with IFRS must mark these statements with iXBRL tags and block-tag notes to such financial statements. Tagging should be performed based on the ESEF taxonomy, which was developed based on IFRS taxonomy. The combination of the xHTML format with iXBRL tags makes reporting suitable for both human and machine reading simultaneously. This approach makes it possible to increase the availability, analysis, and comparability of information obtained in reports.

Artificial intelligence tools and tools for content analysis may as well boost the analysis of the unstructured public data of the companies.

Out of the most discussed instruments, OpenAI and ChatGPT can offer several benefits in analyzing ESG data. Firstly, these advanced AI technologies can efficiently process large volumes of data, enabling organizations to quickly identify and analyze ESG risks and opportunities. Secondly, the natural language processing capabilities of ChatGPT can facilitate the analysis of unstructured data sources, such as social media and news articles, to extract valuable insights that may not be immediately apparent through traditional analysis methods. Additionally, these technologies can assist in the development of ESG policies and reporting frameworks and can support engagement efforts with stakeholders. Finally, the use of AI in ESG analysis can improve accuracy, increase efficiency, and enhance transparency, ultimately contributing to better decision-making and long-term sustainability for organizations.

Modern ESG platforms like ESG Analytics, ESG Book, Embedding Project, and PivotGoals start to use AI technology and big data analytics in general. Platforms use a wide range of sources, often without reference to the company's participation: collection of public documents, websites, mass media, self-registration data, and surveys.

To streamline the process of data analysis, automation is often implemented using various methods. For instance, natural language processing (NLP) is frequently utilized for vast amounts of repetitive data or to conduct sentiment analysis. The RepRisk platform employs machine-learning techniques to scan over 500,000 documents every day in order to identify ESG risks. Furthermore, data augmentation and tensor techniques are employed to complete any gaps in the data, while parsing is utilized to classify and create predictive models, as well as to filter out extraneous data.

Modern data platforms also employ various data presentation methods to make data most suitable for analysis. These methods include integration with Python to build models, applications, and analytics for trading on the stock exchange, as well as tables with historical data that can be outputted in different criteria and forms of graphs. These platforms also provide various data output formats, such as CSV, XML, JSON, WEB, and open API, along with daily data updates, notifications, news digest, and messenger services. Additionally, tools like the ability to search by region, industry, involvement in conflicts, compliance with norms, support for political figures, and company ratings make it easier to work with data.

References

Adams, C. A., & Abhayawansab, S. (2022). Connecting the COVID-19 pandemic environmental, social and governance (ESG) investing and calls for 'harmonisation' of sutainability reporting. *Critical Perspectives on Accounting, 82*, 1–13. https://doi.org/10.1016/j.cpa.2021.102309

Ahmad, K. (2017). The implementation of management accounting practices and its relationship with performance in small and medium enterprises. *International Review of Management and Marketing, 7*, 342–353.

Arcari, A. M. (2018). Preventing crises and managing turnaround processes in SMEs. The role of economic measurement tools. *Management Control, 3*, 131–155. https://doi.org/10.3280/MACO2018-003007

Arumona, J., Erin, O., Onmonya, L., & Omotayo, V. (2019). Board financial education and firm performance: Evidence from the healthcare sector in Nigeria. *Academy of Strategic Management Journal, 18*(4), 1–18.

Aureli, S., Del Baldo, M., Lombardi, R., & Nappo, F. (2020). Nonfinancial reporting regulation and challenges in sustainability disclosure and corporate governance practices. *Business Strategy and the Environment, 29*, 2392–2403. https://doi.org/10.1002/bse.2509

Bank of Italy. (2018, January). Productivity growth in Italy: A tale of a slow-motion change, Occasional Paper n. 422.

Bhattacharya, C. B., & Sen, S. (2004). Doing better at doing good: When, why, and how consumers respond to corporate social initiatives. *California Management Review, 47*(1), 9–24. https://doi.org/10.2307/41166284

Bowen, F. E., & Aragon-Correa, J. A. (2014). Greenwashing in corporate environmentalism research and practice: The importance of what we say and do. *Organization and Environment, 27*(2), 107–112. https://doi.org/10.1177/1086026614537078

Calace, D. (2014). Non-financial reporting in Italian SMEs: An exploratory study on strategic and cultural motivations. *International Journal of Business Administration, 5*(3), 34–48.

Caputo, F., Pizzi, S., Ligorio, L., & Leopizzi, R. (2021). Enhancing environmental information transparency through corporate social responsibility reporting regulation. *Business Strategy and the Environment, 30*, 1–15. https://doi.org/10.1002/bse.2814

Cardoni, A., Kiseleva, E., & Lombardi, R. (2020). A sustainable governance model to prevent corporate corruption: Integrating anticorruption practices, corporate strategy and business processes. *Business Strategy and the Environment, 29*(3), 1173–1185. https://doi.org/10.1002/bse.2424

Cardoni, A., & Paradisi, A. (2020). The implementation of management accounting in small-medium enterprises (SMEs). A knowledge transfer perspective, management. *Control, 1*(Suppl), 39–62. https://doi.org/10.3280/MACO2020-001-S1004

Castellano, N., & Felden, C. (2021). Management control Systems for Sustainability and Sustainability of management control systems. *Management Control, 2*, 5–11. https://doi.org/10.3280/MACO2021-002001

China Securities Regulatory Commission. (2019). *Code of corporate governance for listed companies*. http://www.csrc.gov.cn/csrc_en/c102034/c1372459/content.shtml.

Cho, Y. N., & Baskin, E. (2018). It's a match when green meets healthy in sustainability labeling. *Journal of Business Research, 86*, 119–129. https://doi.org/10.1016/j.jbusres.2018.01.050

Cormier, D., & Magnan, M. (2015). The economic relevance of environmental disclosure and its impact on corporate legitimacy: An empirical investigation. *Business Strategy and the Environment, 24*, 431–450. https://doi.org/10.1002/bse.1829

Dienes, D., Sassen, R., & Fischer, J. (2016). What are the drivers of sustainability reporting? A systematic review. *Sustainability Accounting, Management and Policy Journal, 7*(2), 154–189. https://doi.org/10.1108/SAMPJ-08-2014-0050

Du, X. (2014). How the market values greenwashing? Evidence from China. *Journal of Business Ethics, 128*(3), 547–574.

Eggers, F. (2020). Masters of disasters? Challenges and opportunities for SMEs in times of crisis. *Journal of Business Research, 116*, 199–208. https://doi.org/10.1016/j.jbusres.2020.05.025

Erin, O., Adegboye, A., & Bamigboye, O. A. (2021). Corporate governance and sustainability reporting quality: Evidence from Nigeria. *Sustainability Accounting, Management and Policy Journal, 13*(3), 680–707.

EY. (2020, July). *Study on directors' duties and sustainable corporate governance*. Available at https://op.europa.eu/en/publication-detail/-/publication/e47928a2-d20b-11ea-adf7-01aa75ed71a1/language-en/format-PDF.

Fasan, M., & Mio, C. (2016). Fostering stakeholder engagement: The role of materiality disclosure in integrated reporting. *Business Strategy and the Environment, 26*, 288–305. https://doi.org/10.1002/bse.1917

Ferrarini, G., Siri, M., & Zhu, S. (2021). The EU sustainable governance consultation and the missing link to soft law, Law Warking Paper n. 576/2021. https://doi.org/10.3390/ani11102913.

Gjergji, R., Vena, L., Sciascia, S., & Cortesi, A. (2020). The effects of environmental, social and governance disclosure on the cost of capital in small and medium enterprises: The role of family business status. *Business Strategy and the Environment, 30*, 683–693. https://doi.org/10.1002/bse.2647

Hahn, R., & Kühnen, M. (2013). Determinants of sustainability reporting: A review of results, trends, theory, and opportunities in an expanding field of research. *Journal of Cleaner Production, 59*, 5–21. https://doi.org/10.1016/J.JCLEPRO.2013.07.005

Handley, K., & Molloy, C. (2022). SME corporate governance: A literature review of informal mechanisms for governance. *Meditari Accountancy Research, 30*(7), 310–333. https://doi.org/10.1108/MEDAR-06-2021-1321

Heinicke, A. (2018). Performance measurement systems in small and medium-sized enterprises and family firms: A systematic literature review. *Journal of Management Control, 28*, 457–502. https://doi.org/10.1007/s00187-017-0254-9

Hobbs, A. (2023). How corporate governance can help build a more sustainable world, reporting the European Corporate Governance Conference 2022, Ernst and Young in collabvoration with the Czech Institute of Directors, Association of Chartered Certified Accountants (ACCA), Business Europe and European Issuers, available in https://go.ey.com/3Z6EqM7. https://doi.org/10.1002/chem.202301621

Hristov, I., Chirico, A., & Ranalli, F. (2022). Corporate strategies oriented towards sustainable governance: Advantages, managerial practices and main challenges. *Journal of Management and Governance, 26*, 75–97. https://doi.org/10.1007/s10997-021-09581-x

IFRS - Exposure Draft and comment letters: General Sustainability-related Disclosures. (2022, August). https://www.ifrs.org/projects/work-plan/general-sustainability-related-disclosures/exposure-draft-and-comment-letters/

International Federation of Accountants (IFAC). (2008). *International good practice guidance principles for effective business reporting process*. https://www.ifac.org/system/files/publications/files/Principles-for-Effective-Business-Reporting-Processes.pdf.

Ioannou, I., & Serafeim, G. (2019). Corporate sustainability: A strategy? (SSRN Scholarly Paper No. 3312191). https://doi.org/10.2139/ssrn.3312191

Karim, K., Pinsker, R., & Robin, A. (2013). Firm size and the voluntary disclosure of nonfinancial information by private versus public firm managers. *Managerial Auditing Journal, 28*(9), 866–892. https://doi.org/10.1108/MAJ-01-2013-0800

Kirk, M., & Vincent, J. (2014). Professional investor relations within the firm. *The Accounting Review, 89*(4), 1421–1452. https://doi.org/10.2308/accr-50724

Lai, A., & Stacchezzini, R. (2021). Organisational and professional challenges amid the evolution of sustainability reporting: A theoretical framework and an agenda for future research. *Meditari Accountancy Research, 29*(3), 405–429. https://doi.org/10.1108/MEDAR-02-2021-1199

Lavia Lopez O., & Hiebl, M. R. (2015). Management accounting in small and medium sized enterprises: Current knowledge and avenues for further research. *Journal of Management Accounting Research, 27*, 81–119. https://doi.org/10.2308/jmar-50915

Lyon, T. P., & Montgomery, A. W. (2013). Tweetjacked: The impact of social media on corporate greenwash. *Journal of Business Ethics, 118*(4), 747–757. https://doi.org/10.1007/s10551-013-1958-x

Manzaneque-Lizano, M., Alfaro-Cortés, E., & Priego la Cruz, A. M. (2019). Stakeholders and long-term sustainability of SMEs. *Who Really Matters in Crisis Contexts, and When, Sustainability, 11*, 6551. https://doi.org/10.3390/su11236551

Maraghini, M. P., & Vitale, G. (2018). Sistemi di controllo a supporto dello sviluppo sostenibile delle PMI del settore vitivinicolo: il caso D'Ambra vini s.r.l. *Management Control, 3*, 111–130. https://doi.org/10.3280/MACO2018-003006

Marquis, C., Toffel, M. W., & Bird, Y. (2016). Scrutiny, norms and selective disclosure: A global study of greenwashing. *Organization Science, 27*(2), 483–504. https://doi.org/10.1287/orsc.2015.1039

Michelon, G., & Parbonetti, A. (2012). The effect of corporate governance on sustainability disclosure. *Journal of Management and Governance, 16*(3), 477–509. https://doi.org/10.1007/s10997-010-9160-3

Odriozola, M. D., & Baraibar-Diez, E. (2017). Is corporate reputation associated with quality of CSR reporting? Evidence from Spain. *Corporate Social Responsibility and Environmental Management, 24*(2), 121–132. https://doi.org/10.1002/csr.1399

Porter, M. E., & Kramer, M. R. (2006). Strategy and society: The link between competitive advantage and corporate social responsibility. *Strategic Direction, 23*(5). https://doi.org/10.1108/sd.2007.05623ead.006

Rinaldi, L. (2019). Accounting for sustainability governance: The enabling role of social and environmental accountability research. *Social and Environmental Accountability Journal, 39*(1), 1–22. https://doi.org/10.1080/0969160X.2019.1578675

Sgrò, F., Palazzi, F., & Ciambotti, M. (2022). Business continuity and planning effectiveness: An empirical analysis of Italian manufacturing SMEs. *Management Control, 3*, 89–108. https://doi.org/10.3280/MACO2022-003005

Sustainable Stock Exchanges Initiative. (2023). *ESG guidance database | sustainable stock exchanges*. https://sseinitiative.org/esg-guidance-database/. https://doi.org/10.1021/acs.biomac.3c00431.

Wang, W., Sun, Z., Zhu, W., Ma, L., Dong, Y., Sun, X., & Wu, F. (2022). How does multi-agent govern corporate greenwashing? A stakeholder engagement perspective from "common" to "collaborative" governance. *Corporate Social Responsibility and Environmental Management, 1–17*, 291. https://doi.org/10.1002/csr.2355

Yu, E. P., Luu, B. V., & Chen, C. H. (2020). Greenwashing in environmental, social and governance disclosures. *Research in International Business and Finance, 52*, 101192. https://doi.org/10.1016/j.ribaf.2020.101192. Elsevier.

MIX
Papier aus verantwortungsvollen Quellen
Paper from responsible sources
FSC® C105338

If you have any concerns about our products,
you can contact us on
ProductSafety@springernature.com

In case Publisher is established outside the EU,
the EU authorized representative is:
Springer Nature Customer Service Center GmbH
Europaplatz 3, 69115 Heidelberg, Germany

Printed by Libri Plureos GmbH
in Hamburg, Germany